Advances in Geographic Information Science

Series Editors

Shivanand Balram, Simon Fraser University, Burnaby, BC, Canada
Suzana Dragicevic, Burnaby, BC, Canada

More information about this series at http://www.springer.com/series/7712

Vasily Popovich · Jean-Claude Thill ·
Manfred Schrenk · Christophe Claramunt
Editors

Information Fusion and Intelligent Geographic Information Systems

Computational and Algorithmic Advances
(IF & IGIS'2019)

Springer

Editors
Vasily Popovich
SPIIRAS Hi-Tech Research and
Development Ltd.
St. Petersburg, Russia

Manfred Schrenk
CORP - Competence Center of Urban
and Regional Planning
Vienna, Austria

Jean-Claude Thill
Department of Geography
and Earth Sciences
University of North Carolina
Charlotte, NC, USA

Christophe Claramunt
BRCM Brest, ENCEP
Institut de Recherche de L'Ecole navale
Crozon, France

ISSN 1867-2434   ISSN 1867-2442   (electronic)
Advances in Geographic Information Science
ISBN 978-3-030-31610-5   ISBN 978-3-030-31608-2   (eBook)
https://doi.org/10.1007/978-3-030-31608-2

This Springer imprint is published by the registered company Springer Nature Switzerland AG
The registered company address is: Gewerbestrasse 11, 6330 Cham, Switzerland

Preface

This book contains a number of original scientific papers that were selected after a peer-review process for presentation at the 9th International Symposium "Information Fusion and Intelligent Geographical Information Systems (IF&IGIS'2019)." The symposium was held from May 22 to 24, 2019, in St. Petersburg, Russia, with a specific focus this year on computation and algorithmic advances in IGIS. A separate part of the symposium was devoted to the discussion of achieved results of the international project CRISALIDE. This symposium was organized by SPIIRAS Hi-Tech Research and Development Office Ltd, St. Petersburg, Russia.

The main goal of the IF&IGIS'2019 symposium was to bring together leading world experts in the field of spatial information integration and intelligent geographical information systems (IGIS) to exchange cutting-edge research ideas and experiences, to discuss perspectives on the fast-paced development of geospatial information theory, methods, and models, to demonstrate the latest achievements in IGIS and for applying these research concepts to real-world use cases. The full papers, selected by the international program committee of IF&IGIS'2019, address fundamentals, models, technologies, and services of IGIS in the geoinformational and maritime research fields including underwater acoustics, logistics, environmental management, as well as other modeling and data-driven matters critical to the effectiveness of information fusion processes and intelligent geographical information systems.

The call for papers for the symposium attracted 20 abstracts from 10 countries; 15 papers were selected at the first step of a blind review process for presentation at the conference. After the second step of the review process, the program committee accepted 13 full papers contributed by authors from 6 countries for presentation and publication in this book. In accordance with subjects of accepted papers, four parts of the book were formed: (1) Advances in Intelligent Geographic Information Systems; (2) IGIS integration with acoustics and remote sensing; (3) IGIS Algorithms and Computation Issues; (4) IGIS for Urban and Land-based Research. Special guests of the symposium were two invited speakers who provided us with high-profile lectures on information fusion and intelligent geographical information systems: Professor Vasily Popovich from the SPIIRAS Hi-Tech Research and

Development Office Ltd, St. Petersburg, Russia, and Pierre Laconte President of the Foundation for the Urban Environment, Brussels, Belgium.

The success of the symposium is, undoubtedly, a result of the combined and dedicated efforts of our partners, organizers, reviewers, and participants. We would like to acknowledge the program committee members for their help with the review process. Our thanks go to all participants and authors of the submitted papers as well.

St. Petersburg, Russia   Vasily Popovich
Brest, France   Christophe Claramunt
Charlotte, USA   Jean-Claude Thill
Vienna, Austria   Manfred Schrenk
May 2019

Contents

Advances in Intelligent Geographic Information Systems

The Concept of Space in Philosophy and in Computer Science 3
Vasily Popovich and Alexander Vitko

**CRISALIDE (City Replicable and Integrated Smart Actions Leading
Innovation to Develop Urban Economies): An Experimental Planning
Process Towards Promotion of Innovative Methods and Tools to Face
Contemporary Urban Issues in EU and Russian Cities** 15
Pietro Elisei, Elena Batunova and Miruna Draghia

From Information-Communication City to Human-Focused City 27
Pierre Laconte

A Density-Based Measure of Port Seaside Space-Time Utilization 55
Behnam Nikparvar and Jean-Claude Thill

IGIS Integration with Acoustics and Remote Sensing

**Modeling of Surveillance Zones for Active Sonars with Account
of Angular Dependence of Target Strength and with Use
of Geographic Information Systems** 73
Vladimir Malyj and Andrey Mikhalchuk

**Analysis of Optical Images of the Sea Surface in the Interests
of Environmental Monitoring** 85
Andrey Grigoriev, Filipp Galiano, Maria Zarukina and Vasily Popovich

**Fusing Classification and Segmentation DCNNs for Road Feature
Mining on Aerial Images** 97
Lele Cao and Xin Pan

IGIS Algorithms and Computation Issues

**Information Technologies: Modern Approach to Evolution
of Methods of Obtaining Knowledge About Controlled Processes** 113
Pavel Volgin

**Model Analysis of Maritime Search Operation Using Geoinformation
Technology** 125
Alexander Prokaev

IGIS for Urban and Land-Based Research

**Context-Driven Tourist Trip Planning Support System: An Approach
and OpenStreetMap-Based Attraction Database Formation** 139
Alexander Smirnov, Alexey Kashevnik, Sergey Mikhailov,
Nikolay Shilov, Daria Orlova, Oleg Gusikhin and Harry Martinez

**Regional Geoinformation Modeling of Ground Access to the Forest
Fires in Russia** 155
Ekaterina Podolskaia, Konstantin Kovganko and Dmitriy Ershov

**Concept of Intelligent Decision-Making Support System for City
Environment Management** 167
Elena Batunova, Tatiana Popovich, Oksana Smirnova
and Sergey Truhachev

**Develop a GIS-Based Context-Aware Sensors Network Deployment
Algorithm to Optimize Sensor Coverage in an Urban Area** 179
Meysam Argany and Mir-Abolfazl Mostafavi

Abbreviations

AI	Artificial intelligence
AIS	Automatic identification system
ASS	Active sonar system
AV	Autonomous vehicles
CA	Course angle
CI	Contextual information
CRF	Conditional random field
CRISALIDE	City replicable and integrated smart actions leading innovation to develop urban economies
CSIA	Community urban space innovative agenda
DCNN	Deep convolutional neural network
ES	Expert system
GIS	Geographic information systems
ICT	Information and communications technologies
IDMS	Innovative development governance scheme
IDMSSCM	Intelligent decision-making support system for city management
IDMT	Innovative decision-making tool
IDS	Innovative development schemes
IGIS	Intelligent geographical information system
IHM	Infinite homogeneous medium
IT	Information technology
KPIs	Key performance indicators
LIM	Layered inhomogeneous medium
MAE	Mean absolute error
NBS	Nature-based solutions
NLP	Natural language processing
OOM	Object-oriented modeling
OSM	OpenStreetMap
PCC	Platform Cooperativism Consortium
PCD	Probability of correct detection

PPPP	Public–private–people partnership
RMSE	Root-mean-squared error
SS	Surface ship
Stderr	Standard errors
SZ	Surveillance zone
TS	Target strength
UIA	Urban innovative actions
VRP	Vehicle routing problem

Advances in Intelligent Geographic Information Systems

The Concept of Space in Philosophy and in Computer Science

Vasily Popovich and Alexander Vitko

Abstract In this paper, we continue the discussion about the phenomenon that incorporates two concepts: geoinformation system (GIS) and space. Discussion of our previous works with various experts in computer science, mathematics, philosophy, system monitoring did not change our opinion on this topic. We still believe that the discourse on this subject needs to be conducted from computer science point of view since the expansion of the topic of discussion will lead to a sharp increase in the amount of submitted material and will result in the loss of clarity of the idea that we are trying to convey to the experts. References to philosophical and mathematical sources have been refined after the discussion of the previous paper published in proceedings of IF&IGIS' 17 symposium. In this paper, we again emphasise the idea that every abstraction like "point" has its own interpretation and can be represented in different dimensions: 1D, 2D, 3D or nD. We also wish to note that we again base our arguments on well-known concepts of "space" and "time" from the basic university course of linear algebra [1]. In this paper, we specify the relation of these concepts with algebraic analogue of "space" concept from GIS ("point" concept and its generalisations). Specific characteristics and physical parameters of basic concepts like "point" and "space", their dynamic transformation and variability play an important role for GIS applications. But even more important is the computability of such characteristics using various methods in the interest of the end-user. In this paper, we provide examples of tasks that illustrate our theoretical foundations as a result of computer modelling. We also provide examples of variants of implementation of basic philosophical and computer ideas based on serial products produced by SPIIRAS-HTR&DO Ltd. for more than a decade. This series of products is called "Aqueduct".

Keywords GIS · Space theory · Monitoring system · Measure concept in GIS · "Aqueduct" system

V. Popovich (✉) · A. Vitko
SPIIRAS Hi Tech Research and Development Office Ltd., St. Petersburg, Russia
e-mail: popovich@oogis.ru

© Springer Nature Switzerland AG 2020
V. Popovich et al. (eds.), *Information Fusion and Intelligent*
Geographic Information Systems, Advances in Geographic Information Science,
https://doi.org/10.1007/978-3-030-31608-2_1

1 Introduction

In various scientific approaches and disciplines, e.g. in philosophy, one can find many definitions of space derived from such basic notions like existence and time [2]. Full spectrum of ideas related to notions such as space and time can be divided into two global directions:

- Substantial—Plato, Democritus, Newton, Descartes. Absolute existential autonomy of space and time that remains for all transformations of objects.
- Relational—Aristotle, Leibnitz, Einstein. Dependence of space and time on relations between objects, mass and velocity (here P. denotes the order of coexistence of objects, and B. denotes the order of relation and the order of events).
- Subjective-idealistic—Berkeley, Avenarius. P. and B. are derived from human capabilities to live through and to order events, to place them one after another.

For Kant, P. and B. are a priori forms of sensory contemplation, eternal categories of scene. In his theory of knowledge, Kant for the first time shifts focus from the object of cognition to the cognitive capabilities of the human himself [3]. Kant noted that two worlds exist: actual world (world of things-in-themselves), this world is unknowable, and the second one, visual world (phenomenal world), this world is knowable.

Fundamental remark made by classic metaphysicist Immanuel Kant states that concepts of space and time have no empirical source; i.e. this understanding cannot be derived directly from experience or directly from physical world. These concepts (space and time) are given, synthetic for human understanding and application in practice and in theory.

As a rule [4], modern scientists roughly fall under two groups: overt opponents and adherers of Kant's idea. We have no intention of getting involved in this debate, and we leave metaphysical problem of space and time beyond our focus. Since we regarding artificial systems united under the title computer science and technologies, for the given subject area there is enough evidence to suggest that Kant's idea is close enough to ideology of specialists in computer science and especially in geoinformation as well as modern means of design and development of large heterogeneous and distributed computer systems.

Mindful of the fact that we regard relative spaces in GIS subject area, it makes sense to address special branch of mathematics that operates notion such as space. In doing so, we must not forget that, despite the elements of applied science in it, geoinformation is fully anthropological and oriented firstly on human as a user, as a rule, in the form of convenient and intuitive interface. This topic is discussed in more detail in [4].

Furthermore, it must be noted that today traditional concepts such as "space", "set", "point" are widely used not only in theoretical research but also in practice, e.g. in programming. And here arise serious contradictions that come into sharp focus in geoinformation and in GIS applications. The notion of "point" is so abstract that it becomes harder and harder to apply it in practice, but at the same time it is necessary to preserve theoretical interpretation with affiliation with one of the types of mathematical objects that are applied to describe a subject area or a process. From an everyday perspective, from "human physical sense" point of view, it is impossible to define the concept of "point" as some universal, separate concept (only for the most simple cases, e.g. abstract for coordinate indication). GIS users along with common people always operate in at least three-dimensional space, unlike the point concept that can be understood as a centre of some local coordinate system only in some exceptional cases. This being said, it is difficult to talk about any stable positivity of this abstraction. Any attempts to apply GIS to solving practical tasks result in operations with a large number of abstractions and relations between them. It slows the development, support and efficiency of GIS operation. It is no secret that in various programming languages, e.g. Java, there is a sufficient amount of complex abstractions like point and line, familiar from prior programming languages. As a rule, these abstractions are used for identical notation of abstractions like "point" in Euclidean space. However, in object-oriented, or rather in object languages, we have "object" as an initial abstraction. The notion "object" cannot be interpreted in a similar way to algebraic notion "set". We are dealing with new concept that is closer to "category" concept [5]. Let us also note that "point" concept cannot be interpreted as a coordinate vector. Moreover, we cannot find analogue of abstract data types or algebraic types other than "object" concept itself. Note: It should be noted that many specialists–programmers believe that "object" concept is a further development of idea of abstract data types (N. Wirth et al.).

Thus, a question arises: What is an abstract notion "space" in GIS? For example, if we take the "point" concept for the basis, seeing that it is intuitively understandable concept for practical application in any space, after generalisation of "point" concept we can understand what space we have defined in GIS. The complexity of analysis of this concept increases from the fact GIS is at the same time a theory, a technology and a practical tool for the end-user, sometimes for the simple user, not a specialist. In the light of that, theory, technology and practice very closely interact and often change places in time in non-natural order (considering that we compare existing tradition in fundamental science, application-oriented technologies and pure practice). In computer science, it has been long indicated that practice-oriented technologies often outpace fundamental theoretical research [6]. It happens due to various reasons. Opportunity to start direct implementation of ideas using computer's functionality allows to "skip" the stage of fundamental research; however, as a matter of fact it is untrue. Stage of fundamental research is not "skipped", it is just not reflected in scientific papers and monographs, since it leads to lost time in idea implementation, and it is unfortunately required by modern business. Plus, an important factor is a problem of competition and many results are intentionally withheld by authors or employers with purpose of overtaking opponents.

As an interim conclusion, it may be noted that spaces in GIS can be defined through a set of formats (data, information and knowledge) that are necessary for realisation of required (by client or end-user) business analytics and logic. On the other hand, space in GIS can be defined (similarly to mathematics) as space with defined business analytics and logic. In the scientific literature [7], GIS is represented as a multidimensional space. Because of this, practically all paradigms of mathematical notion of "space" are either trivial and/or useless in the sense of complexity of interpretation. Nevertheless, it is necessary to clearly understand what we are working with and what are the capabilities (requirements) of interpretation. Without the formal definition of GIS, it is difficult to hope for successful development and implementation of GIS application in the long-term perspective. In this sense, space is a basis for systems and its modules specified (defined) for this space. Models act as blocks (initial) for business logic applications which are created for similar representation, study and analysis of objective and\or abstract reality. Indeed, when GIS is applied for relatively simple and traditional task like cartography and local system monitoring, this problem can be not so evident [8]. But when objective is to create, develop and implement global monitoring system (for example) based on GIS, the situation starts to spiral towards loss of control and the system takes life of its own incomprehensible for both developers and especially for users.

If one addresses subject area such as "data fusion in GIS" that can be considered as one variant of theoretical justification of methodology of global monitoring systems' design, it is easy to see that we are dealing with at least six different non-overlapping spaces (e.g. the idea of JDL model design [9] is shown in [7]). Bearing in mind the fact that, following Kant's hypothesis, concepts such as space and time are artificial (nonobjective), and GIS such as a science, technology and practical application cannot have objective fundament (basis). The most objective data can be considered as measurement data which can later have various interpretations and applications in tasks. It can, therefore, be deduced that any space in GIS is artificial. In practical sense, it means that one must be able to work with external data flow, preferably consisting of measurement data. Consequently, we must clearly understand that there are three principal parameters (categories): measurement time, measurement point and measuring tool. With time, a lot of data lose their actuality or their meaning. That is why categories such as time and space are crucial for GIS and for any system that uses GIS platform.

2   Definition of Space Concept

Having discussed "time" and "space" categories as crucial concepts for GIS, let us note that actuality of this discussion is enhanced by practical application of GIS in global monitoring systems. The concept of time is universal for the whole area of applied science that is related to GIS subject, and it is now difficult to imagine something special related to this concept (time). It is only worth noting that we can have various relative timescales. For example, in modelling of processes or of business logic one is able to slow down or speed up time of execution; however, it has no effect on the time concept itself. Sometimes, one can witness up to five (according to our practice) different timescales going in parallel or in succession. As a rule, such approach is applied in interactive (interacting with human) modelling systems.

Everything is much more complicated with "space" concept. Let us consider the previously noted fact that monitoring systems based on GIS contain six basic types of spaces (scientific subject areas), yet in reality their number can significantly increase and they should be regarded both separately and jointly.

As was mentioned in the preface to this paper, for common user, on the one hand, the most simple notion is "point", but at the same time, "point" is the most abstract concept from scientific and technological points of view. The concept of "point" itself is indeed a primitive concept that includes or points at some set of properties. However, "point" concept is not independent or universal. Without common notion of "space" or its specification in mathematical sense, the concept of "point" is meaningless. Addressing our subject area (GIS), "point" is not only a mathematical abstraction, and, above all, "point" in GIS is associated with several coordinates. At least, it can always be translated into coordinates. But, depending on context, point not only denotes coordinates and usually it has a whole number of properties, sets of characteristics and even methods (functions) to go with it.

For simplicity, let us consider different variants of point definition:

(a) One-dimensional case (point itself). The point has one coordinate but at the same time can have or point at a number of other parameters.
(b) Two-dimensional case (Euclidean space). The point has two coordinates and points at some non-empty of other parameters.
(c) Three-dimensional case. The point has three coordinates and a reference (pointer).
(d) Multidimensional case. The point contains coordinate vector and a reference.
(e) All cases listed above plus time. The point has additional parameter—time.

On the one hand, point is a fundamental concept based on which all other derived abstractions in GIS are formed. On the other hand, "point" abstraction should not be constant and should change according to the type of space. If we again return to JDL model, let us once more note that every level has its own specific space with specified business analytics (structure). Drawing attention to the fact that at this point GIS is applied for specific tasks belonging to one subject area, thus such spaces ought to strictly be non-overlapping, and otherwise one can get the whole system of contradictions.

Let us make another important observation. It is quite difficult to build a GIS system based on the idea of non-overlapping spaces using algebraic axiomatic approach [7]. It means that it is practically impossible to formulate a universal set of notions and definitions (ontology or ontology system) for all types of spaces in GIS. It is thus expected to apply evolutionary (not axiomatic) approach to building a GIS platform. Apparently, the creation of consistent axiomatics and theory is hardly achievable in the nearest future. Our reasoning implies that we can address the system of typical spaces matching the levels of JDL model (Fig. 1). For example, let us consider the system of applied spaces implemented in maritime monitoring system "Aqueduct" produced by SPIIRAS-HTR&DO Ltd.:

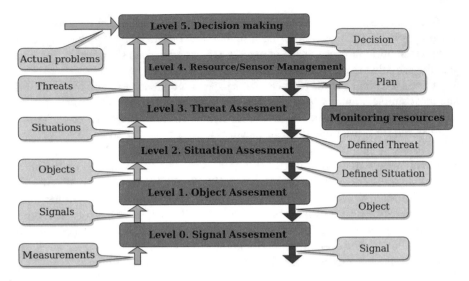

Fig. 1 JDL model for "Aqueduct"

(a) Zero or first level, space of measurements in one or several physical spaces (atmosphere, ocean, etc.).
(b) Second level, space of objects and their tracks.
(c) Third level, space of tactical situations (TSs). Tactical situation is a time sample of a state of the full range of spaces (subspaces): environment (atmosphere, ocean, etc.), available resources, challenges or problems, goals to be achieved. As a rule, TS space is specified beforehand and is based on scientific and/or expert knowledge, but it can be dynamically formed and/or added to using various mathematical and computer techniques.
(d) Fourth level, space of threats, challenges and hazards with their quantitative (preferably) assessment.
(e) Fifth level, space of available resources that can be used and applied.
(f) Decision space as a reaction to space of threats. It is also specified beforehand in the form of typical scenarios and templates; however, it can be crated dynamically in case of emergence of untypical threats and upon decision-maker's (DM) demand.

The example of maritime monitoring system illustrates that for the given subject area there exists a rather complex combination of various spaces that are closely interconnected with each other but have a fundamental difference which is a notion of "point in space" as initial abstraction with regard to GIS [10]. In short, the "point" abstraction can be specified as a measurement (of a signal), an object (physical or abstract), a tactical situation, threat, resource and decision [11]. Bearing in mind the fact that application aspect of research is of the most importance and interest for geoinformatics, a valid aspect arises regarding the concept of "measure" as the basis for creation of business analytics.

3 Specification of Measure on a Space

As was noted before [4], measure in philosophical aspect is a philosophical category and indicates uniqueness of qualitative and quantitative properties of some object. However, analysis of measure should be performed while taking into account the time interval in which the measured values are saved. "Measure" category is related to a certain number of philosophical definitions connected to ethics and aesthetics. In mathematics, measure is a common notion for various types of generalisations like Euclidean length, area and n-dimensional volume.

There are various specifications to the notion of measure (mathematical point of view):

(a) Jordan measure is an example of finitely additive measure or one of the ways of formalising notions of length, area and n-dimensional volume in Euclidean space.
(b) Lebesgue measure is an example of denumerable additive measure and is a continuation of Jordan measure on more vast class of sets.
(c) Riemann measure (Riemann integral) is an area of region under a curve (a figure between graph of function and abscissa).
(d) Hausdorff measure is a special mathematical measure. Necessity of introduction of such measure derived from the need to calculate length, area and volume of nonspecific figures that can be not specified analytically.

Along with the "usual" concept of "measure", application of Hausdorff measure in GIS opens new opportunities to expand business analytics in GIS applications and particularly in complex, multidimensional spaces. As an example, one may take calculation of a volume of acoustic field in two-dimensional (Fig. 2) and three-dimensional (Fig. 3) representation. This task cannot be solved analytically since it cannot be represented as one of the known analytical functions. There exists only a set of algorithms that executes this particular calculation.

Quite possibly, the same level of complexity occurs when calculating the volume of network filed coverage (3G, 4G, Wi-Fi, etc.) for the end-user with consideration of his real location in the environment (buildings, metal plating, etc.). Numerical values of such fields in space can be calculated (often only theoretically) if one knows environment's transfer function (Green's function). To do so, one must specify the Hausdorff measure and apply step-by-step calculation. It is possible to apply simulation modelling when the field is already known in the form of some data set.

Similar tasks are common in radiolocation. Our last research illustrates that radio wave propagation medium has almost the same stratification as sea water. Moreover, the period of changes and the scale of these changes are even more considerable in the atmosphere than in the ocean. Calculation of radar field for heterogeneous atmosphere is shown in Fig. 3.

All calculations shown in Figs. 2, 3 and 4 illustrate one major difficulty. A unified function or a system of such functions for execution of these calculations does not exist. It means that no inversion can be made for this calculation task. And for

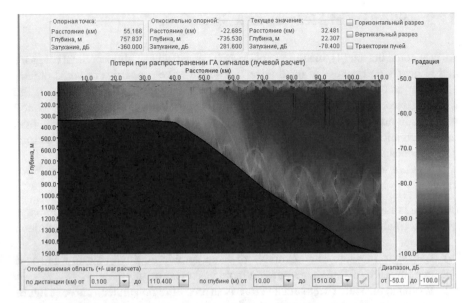

Fig. 2 Example of calculation of two-dimensional acoustic field

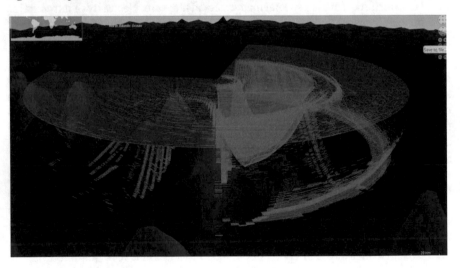

Fig. 3 Three-dimensional model of the acoustic field of acoustic source in the ocean

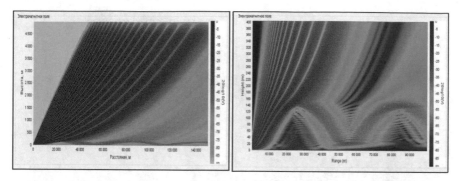

Fig. 4 Calculation of radar field

derived calculations that are required in practice, one has got only a matrix of values. Moreover, these matrices are of an enormous size. In this regard, Hausdorff measure is a very appropriate concept for execution of derived calculations for such fields.

4 Case Study

In this section, let us give the examples of visual representation of various types of spaces that are logical spaces and can be perceived by humans as one universal space (visually). And herein lies a misconception, when a developer thinks the same way as a user and tries to develop a system not from scientific and technological point of view but from the point of view of a user. In other words, development progresses from the interface.

Figure 5 illustrates several independent applied spaces: cartographic data in Open-StreetMap format and space of object information about vessels at sea and space of object information about aircraft as presented in the "Aqueduct" system. Common for all given types of spaces are coordinate system and time, i.e. some basis that serves to link vertically and horizontally all spaces in GIS. At the same time, the distinctive features of all spaces are their own notions of point and their own analytics. Also, a separate ontology can be designed for each space.

It is worth noting that some analytics can generate new types of spaces, some of which can have visual representation suitable for humans. To give an example, let us refer to a classifier based on singular value decomposition of a matrix, SVD method. This method obtained applied significance through scientific works of Prof. Tarakanov A. O. It was first applied in description of operation of immune system and was called "immunecomputing" [12].

This method has shown its universality and is widely applied in "Aqueduct" products as a part of artificial intelligence subsystem. An example of application of SVD method for reconstruction of hydrophysical field of the Black Sea is illustrated in Fig. 6.

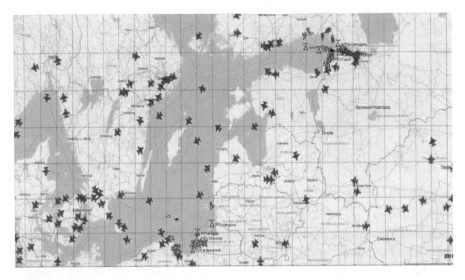

Fig. 5 Composition of various spaces in GIS as one image

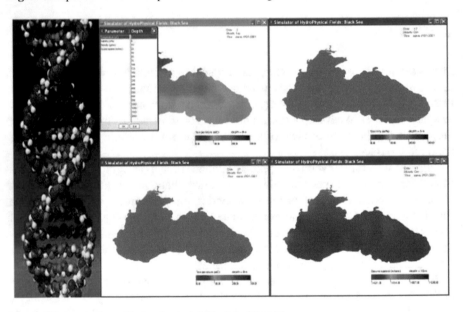

Fig. 6 Reconstruction of hydrophysical field of the Black Sea

The major point of the example is explained below. One is given an initial database of measurements (data for the last hundred years) of such fields as water temperature, salinity, conductivity, speed of sound on various horizons on different days and in different seasons. In other words, data were obtained from various measurement stations and expeditions. Initial database was processed by SVD machine and, as a result, statistical average year with twenty-four-hour discrete step was obtained. A comparison of modelling results with current measurements has proven good statistical stability of the model and has shown compliance with all major hydrophysical processes that occur in the Black Sea. The obtained results can be presented in the form of some specific space that can interact with other types of spaces, or can be used as initial data for various types of analytics.

5 Conclusion

Abstract concept of "space" not only has its own theoretical significance in GIS research but also is an example of successful application of theoretical results in a wide range of subject areas.

At first sight, it might appear that the concept of "space" in GIS is close to the "layer" concept. Although these concepts visually seem similar, nevertheless, they have the following fundamental differences:

- "Layer" concept is a set of attribute information prepared for display on the computer screen. "Space" is a mathematical structure with a specified axiomatics that can as well be visually displayed on the computer screen in the form of a set of attribute information.
- "Layer" is a static display element, i.e. a "snapshot".
- "Space" can generate "layers", and the converse is not true.

The need to introduce the concept of "space" resulted from sharp increase in complexity of tasks solved by GIS. It is not only the problem of big data, but above all the problem of "big analytics" and of a blend of rater complex subject areas. For instance, in "Aqueduct" products we implement mathematical physics and its applied aspects like underwater acoustics, radiolocation and hydrophysics.

We are also forced to implement artificial intelligence systems to solve tasks that either cannot be solved or are difficult or economically impractical to solve using traditional mathematical and statistic methods.

Critical issues pose operating systems and hardware architecture, especially when GIS becomes the core of real-time systems. At the present time, there are no suitable ready-made operating systems and hardware architecture products. All solutions that exist on the market need serious refinement and configuration for each complex subject area.

Justification of suitable operating systems and hardware architecture that can implement soft, hard and hybrid real-time modes will be the main subject of our further research.

References

1. Shraer O, Shperner G (1934) Introduction to linear algebra in geometrical interpretation (Einfuhrung in die analytische Geometrie und Algebra). Olshansky G (translation from Germany to Russian). ОНТИ M-L, 210 p
2. Ogurtsov AP (1988) Disciplinary structure of science. Genesis and argumentation (Russian)
3. Kant I (1965) Critique of pure reason (Kritik der reinen Vernunft) [1781] (trans: Norman Kemp Smith), A 51/B 75. St. Martins, N.Y.
4. Popovich V (2017) Space theory for IGIS. Information fusion and intelligent geographic information systems. New frontiers in information fusion and intelligent GIS: from maritime to land-based research. Shanghai Maritime University, China, May 2017. Lecture notes in geoinformation and cartography. Springer, Berlin, Heidelberg, pp 3–12
5. MacLane S (1971) Categories for the working mathematician. Springer
6. Eilenberg S, MacLane S (1945) A general theory of natural equivalences. Trans Am Math Soc 58:231–294
7. Popovich V (ed) (2013) Intelligent GIS. Nauka, Russian, 324 p
8. Thill J-C (2011) Is spatial really that special? A tale of spaces. Information fusion and geographic information systems: towards the digital ocean. Lecture notes in geoinformation and cartography, vol 5. Springer, Berlin, Heidelberg, pp 3–12. https://doi.org/10.1007/978-3-642-19766-6_1
9. Blasch E (2002) Fundamentals of information fusion and applications. Tutorial, TD2, Fusion
10. Popovich V (2016) Detection and search theory for moving objects. Nauka, Russian, 423 p
11. Koopman BO (1956) Theory of search: 2. Target detection. Oper Res 4(5)
12. Tarakanov AO, Skormin VA, Sokolova SP (2003) Immunocomputing. Springer, New York

CRISALIDE (City Replicable and Integrated Smart Actions Leading Innovation to Develop Urban Economies): An Experimental Planning Process Towards Promotion of Innovative Methods and Tools to Face Contemporary Urban Issues in EU and Russian Cities

Pietro Elisei, Elena Batunova and Miruna Draghia

Abstract CRISALIDE is one of the very few projects financed between EU and Russian Federation through the ERA NET-RUS PLUS programme. The change, as the catalyser of innovation, is embedded in urban life. The change covers different fields; for instance, the demography, the citizens' behaviours, the working patterns, the use and extension of both public space and sphere the modes/means of production and so on. Urbanisation is a great opportunity for supporting innovative choices and urban solutions. CRISALIDE aims at experimenting an innovative methodology looking at linking the decision to different factors both connected to horizontal and vertical governance. A participatory planning approach concretely looks for a multi-level governance dialogue and relies on state-of-the-art technology to design an e-platform facilitating the decision-making process. The platform is abstracting, digitalising and finally creating a replicable and user-friendly tool based on an enlarged participatory planning process grounded on public–private–people partnership (PPPP) principle. The platform is harmonising the contribution of stakeholders in diverse planning domains and formalise them through key performance indicators (KPIs) providing values at disposal to decision makers linked to grade of smartness and comprehensive quality of life generated by the triggered regenerative/transformative planning process. Strategic, smart and integrated urban management is a key tool to promote stable growth and effective processes of innovation. New efforts to modernise the

P. Elisei (✉) · M. Draghia
Urbasofia, Bucharest, Romania
e-mail: pietro.elisei@urbansofia.eu

E. Batunova
Southern Urban Planning Center, Rostov-on-Don, Russia

M. Draghia
Ion Mincu University of Architecture and Urban Planning, Bucharest, Romania

© Springer Nature Switzerland AG 2020
V. Popovich et al. (eds.), *Information Fusion and Intelligent Geographic Information Systems*, Advances in Geographic Information Science,
https://doi.org/10.1007/978-3-030-31608-2_2

Russian economy and face global issues as well have taken on an even greater significance since the implementation of Western sanctions. Cities can be the natural catalyser for promoting innovation, as they contain all strategic elements, at a scale of proximity. This paper will show the state of play on discussions related to urban planning and innovation; it will focus on the Russian case and provide evidence of first achieved results in the testbed area situated in Rostov-on-Don.

1   Innovation and Urban Planning

Innovation in urban planning covers different fields; for instance, the demography, the citizens' behaviours, the working patterns, the use and extension of both public space and sphere the modes/means of production and so on. In the European context, especially of urban policies promoted by the EU, in the last few years, the theme of urban innovation has become a mantra that supports various direct financing programmes and indirect ones. The Great Financial Crisis of 2007 created a new situation: the severe changes in economic and financial conditions extorted the revision of existing public policies. Under the worsening financial conditions, the traditional functioning of cities collapsed: urban regeneration projects stopped, social welfare systems have been curtailed and private market investments were postponed [23]. In 1950, 30% of the world's population was urban, and by 2050, 68% of the world's population is projected to be urban. Under these new conditions, urban planning had to become more flexible and more open to unusual ideas and notions coming from the local communities. Parallel to the weakening of the old-style political machinery and the innovative changes in municipal policies, in many countries, citizens became much more active and strived for more influence over urban development.

Their political will towards more sustainable, equitable, inclusive local societies can only be approached by new public policies based on openness towards the bottom, the citizens and NGO-s, as the upper political levels are usually against these ideas. The world's top 100 cities will account for 35% of global GDP growth between now and 2025 (McKinsey Report 2011). Urban settings magnify global threats such as climate change, water and food security and resource shortages, but also provide a framework for addressing them. If the future of cities cannot be one of unsustainable expansion, it should rather be one of tireless innovation. This report chronicles ten of the best examples from around the world of how cities are creating innovative solutions to a variety of problems [5]. Many of these solutions are scalable, replicable and can be adapted to a variety of specific urban environments. From smart traffic lights to garbage taxes, innovations in technology, services and governance are not ends in themselves but mean to shape the behaviour and improve the lives of the city's inhabitants. All innovations should be centred on the citizen, adhering to the principles of universal design and usable by people of all ages and abilities. It is vital to look with different perspective to

- the underused resources available in the urban environment (initiating temporary uses, rethinking use of public spaces, introducing practices of circular economy, optimising use of energy at neighbourhood scale, promoting nature-based solutions...);
- the way we move and access goods and services in the urban environment (promoting multi- and inter-modal concepts for mobility, creating green infrastructures favouring low impact mobility means such as bicycles, promoting the implementation of location-based services...);
- enlarging the decision-making processes in order to add creativity, but at the same time concreteness, in defining the development target at urban scale: citizens, stakeholders, city users have to be involved in the design of the city together through a cooperativistic approach. Sustainable development is based on local rooted solutions, rooted in the real life of local communities.

In Europe, the Innovation Union Programme was launched in 2010 as a flagship initiative of the Europe 2020 strategy to build on Europe's strengths and address its weaknesses with respect to innovation and thereby make Europe more competitive in times of budgetary constraints, demographic change and increased global competition. The main achievements under this priority are the implementation of the European Research Area and the launch of Horizon 2020, the new research and innovation framework programme, streamlining funding and encouraging cross border and transnational collaboration.

The flagship programme of the EU addressing the cities is the Urban Innovative Actions (UIA). Cities are going to be financed after proposing innovative ideas co-designed together with a local partnership constituted by key stakeholders, basically following the quadruple helix principle. Urban Innovative Actions test innovative ideas and support urban authorities in their efforts to ensure sustainable urban development. In 2015, urban stakeholders and member states identified 12 topics which represent common challenges cities are facing: (1) air quality, (2) innovation and responsible public procurement, (3) circular economy, (4) integration of migrants and refugees, (5) climate adaptation, (6) jobs and skills in the local economy, (7) digital transition, (8) housing, (9) energy transition, (10) sustainable use of land (nature-based solutions), (11) urban mobility, (12) urban poverty (deprived neighbourhoods).

All over Europe, there are currently 54 projects experimenting with urban innovative solutions for current issues faced by EU challenges (https://uia-initiative.eu/en/uia-cities-map) (Fig. 1).

Fig. 1 Innovative actions currently financed all over EU

2 Experience Innovation in the Russian Urban Context

Despite the fact that the country' s transition to the new socio-economic model is associated with the introduction of the unknown for the Soviet model conditions, such as private property and private actors and the creation of the new urban planning system, this system in many aspects replicates the obsolete socialist model of planning. It is still top-down, centralised, comprehensive and does not consider public participation as an important element of decision making. In fact, even with the big role of Russia's science and its potential, today the country seems to be lagging behind in cultivating high-tech technologies for innovating the economy of cities. Although it is one of the leaders in developing space, defence and nuclear technologies, Russia is drastically falling behind in producing consumer technologies [16]. Innovations in the contemporary Russian urban planning system are mainly seen in the usage of the new technological tools, such as GIS, but the urban planning system is far from the introduction of organisational or social innovations.

A remarkable asset of CRISALIDE lies in the fact that it is necessary to arrive to the definition of a platform, based on specific urban and territorial indicators, which facilitates decision-making choices in urban transformation and regeneration processes. In this sense, both the European and the Russian partners are bound by the concreteness of the real problems connected to the transformation of an urban area. Moreover, this platform is designed through making a real experience of participatory planning, which is looking at innovative governance links, both horizontally (the local level) and vertically (the involvement of the different governance layers, the multi-level governance approach). The project's name is already a synthesis of the project attitude towards the application of innovative methodologies in the urban realm. The chrysalis (CRISALIDE is the Italian version of the word chrysalis) is a butterfly at the stage of development when it is covered by a hard case before it becomes an adult insect. It is a metaphor of the translation towards a new state of play characterised by the application of innovative solutions, as innovation is the result of a process that transforms the state of things bringing them to a completely new situation: the chrysalis becoming a butterfly.

CRISALIDE through its methodology and participatory activities is developing a stable, engaged stakeholder group, which will assist in the establishment of the innovative development schemes (IDS) tackling aspects related to housing, mobility and infrastructure, public space and services/facilities, environment, landscape and heritage, as well as urban management and governance. The results and impacts are organised in three innovation areas:

- organizational innovation (such as new niches for local, city-based private sector to boost R&D and innovation activities, policy impact to reinforce local and national related policies regarding collaboration in the field of R&D and innovation),
- technological innovation and
- social innovation (enhanced local identity to improve social capital, increased climate and environmental awareness to favour community's preparedness, increased ICT development awareness to enhance local economics).

CRISALIDE project is experimenting a joint EU-Russian research and collaborative approach for the creation of a digital innovative platform, designed to facilitate the renewal and regeneration of abandoned areas and brownfields. As an outcome, the platform is envisioned to become a replicable, user-friendly tool, aligning and harmonising visions from a diverse ecosystem of stakeholders. One of the main expected outcomes of the participatory-based methodology is to boost, strengthen and consolidate the collaboration between multiple stakeholders towards RDI in the city of Rostov-on-Don. As a natural consequence, an innovative decision-making tool (IDMT, the envisaged e-platform) will be defined in a participative environment and setting, following the core principles of stakeholders' empowerment.

The IDMT tool is an intelligent geographical information system (IGIS) based online platform, which aims to bring together researchers, technology providers (local, national and international as well) for R&D and innovation. The tool will be developed (at least) on four main innovative development schemes (IDS) pillars (planning domains), developed together with sectorial key stakeholders facing the main domains ruling the function of urban areas (area-based approach) and defining for them potential pathway of/for innovation in products and services. Participatory processes are widely known practices in city planning and development projects. Several methods and practices can achieve certain levels of citizen/stakeholder involvement. In this context, CRISALIDE project focuses mainly on consultation and collaboration methods, in order to achieve long-term collaboration in the field of RDI among researchers, businesses and companies and the public sector. The engagement and active participation of different local actors is crucial for reaching a successful design and conception of the IDMT.

3   The Ongoing Trial in Rostov-on-Don

The conceptualisation of the participatory-based methodology is embedded into a theoretical background on participatory planning processes and design thinking methods, in order to design the most suitable workshop scenario for identifying local needs, problems, opportunities and brainstorming potential solutions to the most pressing challenges. One of the main expected outcomes of the participatory-based methodology is to boost, strengthen and consolidate the collaboration between multiple stakeholders towards RDI in the city of Rostov-on-Don. By applying a participatory-based methodology and approach, the idea of empowerment will provide stakeholders with the ability and capacity to become agents for change in the process of decision making concerning their own lifestyles and environment [2]. As a natural consequence, the innovative decision-making tool (IDMT) will be defined in a participative environment and setting, following the core principles of stakeholders' empowerment. The IDMT is the technological face of a governance scheme defined and experimented together with the local stakeholders both working at local or at other governance levels; this is the innovative development governance scheme (IDMS) (Fig. 2).

The development pathway agreed through the IDMS will provide the inputs to set up the IDMT. The combination of scheme and derived tools will facilitate the definition of an evidence-based urban/territorial strategic development agenda for the selected intervention area: the community urban space innovative agenda (CSIA).

Going beyond the narrative of participatory approaches and stakeholders' empowerment in the process of planning and decision making, another key challenge lies in the gap between the ideology-based policy making and the more pragmatic evidence-grounded decision-making process [7]. According to Davies' understanding, the evidence-based policy and decision making represent an approach that "helps people

---

CRISALIDE is trying to define an innovative approach through:

   a)   A Ground-breaking Methodology in Urban Planning (IDMS)
   b)   Scalable & Replicable Online Collaboration Platform (IDMT)

IDMS= Innovative Decision Making Scheme
IDMT= Innovative Decision Making Tool

IDMS + IDMT → leads to the **CSIAs (C**ommunity urban **S**pace **I**nnovative **A**gendas**)**

Fig. 2   Essentials of the CRISALIDE approach

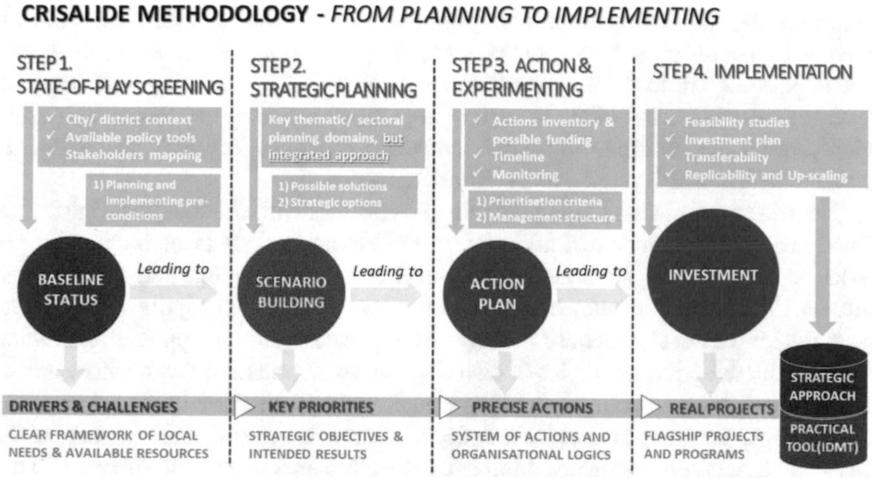

CRISALIDE METHODOLOGY - *FROM PLANNING TO IMPLEMENTING*

STEP 1. STATE-OF-PLAY SCREENING	STEP 2. STRATEGIC PLANNING	STEP 3. ACTION & EXPERIMENTING	STEP 4. IMPLEMENTATION
✓ City/ district context ✓ Available policy tools ✓ Stakeholders mapping	Key thematic/ sectoral planning domains, but integrated approach	✓ Actions inventory & possible funding ✓ Timeline ✓ Monitoring	✓ Feasibility studies ✓ Investment plan ✓ Transferability ✓ Replicability and Up-scaling
1) Planning and implementing pre- conditions	1) Possible solutions 2) Strategic options	1) Prioritisation criteria 2) Management structure	
BASELINE STATUS	*Leading to* → **SCENARIO BUILDING**	*Leading to* → **ACTION PLAN**	*Leading to* → **INVESTMENT**
			STRATEGIC APPROACH
DRIVERS & CHALLENGES	KEY PRIORITIES	PRECISE ACTIONS	REAL PROJECTS	PRACTICAL TOOL(IDMT)
CLEAR FRAMEWORK OF LOCAL NEEDS & AVAILABLE RESOURCES	STRATEGIC OBJECTIVES & INTENDED RESULTS	SYSTEM OF ACTIONS AND ORGANISATIONAL LOGICS	FLAGSHIP PROJECTS AND PROGRAMS	

Fig. 3 CRISALIDE methodological approach

make well-informed decisions about policies, programmes and projects by putting the best available evidence from research at the heart of policy development and implementation". Therefore, it is evident that the IDMT should be founded on sound reasoning processes and backed by reliable and actual information and data, leveraging on the evidence-based planning as a rigour and systematic approach to achieve replicable and scalable results [9] (Fig. 3).

Brownfield redevelopment in Russia is a quite new and challenging issue of urban planning. As such, planners' attention to it has just started to arise, which is being explained by several factors and determinants. Firstly, compared to the highly urbanised territories of the European countries, where the issue has become one of the EU policy's priorities [19], Russia has vast greenfield areas near the cities, whose development requires significantly less efforts in terms of economy and legislation. Similar situation is observed in the USA, where the interest to brownfield redevelopment is also much lower than in the EU [17]. Secondly, the traditional socialist way of extensive urban growth was not limited by any economic factors and there were not any incentives to recycle land [3]. The following post-socialist construction boom provoked even higher land consumption and urban sprawl due to the high housing demand and interest from both public and private parties to satisfy the population's needs [20]. However, after three post-socialist decades, the unsustainability of such approach is becoming evident especially in large capital cities, which have been attractive poles for both population migration and investments into housing construction. Traffic, air pollution, lack of social facilities, services and technical infrastructure make new peripheral areas of the large cities less attractive and problematic for the city management. These problems provoked several attempts to reorient urban development inside the existing city territory and consider brownfields as an important development resource, given its intention of appearance in the national policies [1]. Rostov-on-Don provides the ideal case to redevelop a brownfield that is the old airport area. In this area, many interests of developers are already put in place and different development ideas are already circulating. CRISALIDE tests here its approach looking at putting in place an inclusive and sustainable planning process, which tries to propose innovative solutions, attitudes and values in urban transformations related to the Russian context.

The site for implementation was selected together with the local administration among several former industrial areas in Rostov-on-Don. It is recommended for redevelopment by the strategy of socio-economic development and attracts high investors' interests. The choice was due to the value of such a significant territorial resource for the development of the city, great prospects for the application of innovative methods and technologies in planning the development of this territory and its development, the potential for creating a high-quality urban environment, creating opportunities for innovative economic development, forming a positive image of the city and attracting investments. Moreover, the large area of the site (about 350 ha) involves a long-term phased implementation of the project (Fig. 4).

The area of the former airport "Rostov-on-Don" is located in the eastern part of the city in nine kilometres from the city centre, in the Pervomaysky administrative district. The airport stopped its operation on December 2017, when the new international

Fig. 4 The intervention area (old airport) in Rostov-on-Don

airport was opened—Platov International Airport. In the local planning documents, the old airport's territory was considered as an internal spatial resource for development years before the actual realisation. Thus, the City General Plan approved in 2015 proposed the construction of 1596 thousand square metres of housing within the plot of 267 ha until 2035. After the new airport construction, the local authority started to promote the area for redevelopment and several projects have been done, one of which was presented at the Russian Investment Forum in Sochi in 2018. The site is located in the peripheral part of the city, but at the same time in the centre of Rostov-on-Don metropolitan area (named "Big Rostov") that consists of eight cities and a number of rural settlements and whose population number is estimated at about 2.2 million people. The site is surrounded by the important areas of commercial activities of regional significance: the regional markets and shopping malls with such actors as IKEA or Auchan are operating here. The transport axis, on which the site is located, links the city centre with the new airport and the very important federal road M4 Don, which connects the national capital Moscow with the Black Sea coast and countries of the Caucasian region.

CRISALIDE is currently experimenting in Rostov-on-Don, so it has not yet come to define this platform and first outputs are under a critical examination operated by the project's partners and other involved stakeholders, next step is the formalisation of the IDMT. From a strictly urbanistic point of view (there is also in CRISALIDE a translation of qualitative themes into ontologies useful to computer scientists to create the platform, but they are not the main object of this article), the themes (urban functions) on which to work were identified through the definition of the IDMS (basically through the workshops held in Rostov with key stakeholders). Each of these themes is being evaluated based on a set of key performance indicators (KPIs) based on the following urban functions (planning domains):

- The places of living (housing)
- The spaces for production
- The public space
- Mobility and accessibility
- Governance and participation
- Green infrastructure (NBS, nature-based solutions).

Expressed in very simple terms, in reality the platform created is based on ontologies, relationships and algorithms that are far more complex, CRISALIDE tries to combine planning domains, classic urban functions, established through a participatory process, with indicators that guarantee the highest quality of the planification process (e.g. correct use of resources, attention to the use of public spaces, correct use of technologies for services to citizens, attention to climate change, correct use of construction materials and so on).

However, in simple terms, n order to be able to measure and transform a qualitative planning process into a quantitative one, therefore measurable and scalable, a function of several variables is defined, each variable corresponds to a planning domain

$$F(PD1, PD2, PD3, PD4, PD5 \ldots)$$

The initial value of F is providing the state of play of the urban area to transform; this value is going to change with the variation of the KPIs.

For each planning domain, there is an ideal value to achieve.

This ideal value is achieved when every KPI provides an optimum in its range of values.

Basically, the same PD assumes its value depending on the combination of KPIs values

Moreover, not all the KPIs have the same weight (experts/planners define their weight)

$$PD\ (aKPi1, bKPi2, cKPi3, \ldots)$$

The definition of the KPIs is currently under definition in the project.

4 Conclusions

The CRISALIDE project promotes a way of working on the cities that is very distant from the ordinary practices operated in Russia. The CRISALIDE methodology builds the solutions from the bottom, works with the stakeholders in identifying the problems to be faced and defines the figure of the planner as that of a mediator and facilitator of complex processes. A methodology that has no a priori solutions, which does not have the innovation package ready to be sold and applied top down. The digital dimension is an arrival point of a real participatory planning process.

The CRISALIDE project being oriented to the development of a digital innovative platform creates an opportunity to introduce social and organizational innovations in urban planning through participatory bottom-up approach. The digital dimension is an arrival point of a real participatory planning process. The platform, as a tool that facilitates balanced planning and as a tool that facilitates sustainable decision-making, is born of an advanced dialogic context (IDMS, innovative decision-making scheme). Reducing the complexity of a complex process to an IT tool means losing many nuances, but the advantage of the digital tool (IDMT) is to focus on the essential elements of the planning process, in terms of urban functions and related indicators.

This methodology aims at

- avoiding the construction of commercial housing,
- promoting the formation of high-comfort public spaces,
- supporting the introduction of new forms of mobility and environmentally friendly technologies,
- backing up the creation of conditions for maintaining the health of the population,
- facilitating the realisation of creative and intellectual potential,
- triggering the activation of innovative forms of economies.

Acknowledgements This paper is based on the project "CRISALIDE: City Replicable and Integrated Smart Actions Leading Innovation to Develop Urban Economies", the winner of a joint

Russian-European competition of innovative projects under the ERA.NET-RUS PLUS programme (project no. 42016, application ERA-RUS-4188).
The implementation of the CRISALIDE innovative project is funded by the Executive Agency for Higher Education, Research, Development and Innovation Funding (UEFISCDI), Romania [grant number CTR 69/2018]; the Foundation for Assistance to Small Innovative Enterprises in Science and Technology (FASIE), Russia [grant number 395ГР/42016].

References

1. Batunova E, Trukhachev S (2019) Searching for smart solutions in urban development beyond the political slogans: a case of Rostov-on-Don, Southern Russia. REAL CORP conference proceedings, ISSN 2521–3938. Available at https://archive.corp.at/cdrom2019/papers2019/CORP2019_121.pdf
2. Benson J, Roe M (2007) Landscape and sustainability, Second edn. Routledge Taylor & Francis Group, London and New York
3. Bertaud A, Renaud B (1997) Socialist cities without land markets. J Urban Econ 41(1):137–151. https://doi.org/10.1006/juec.1996.1097
4. Bisello A, Elisei P, Stephens R, Vettorato D (2015) Smart and sustainable planning for cities and regions. Springer
5. Charles A () These are the top 10 urban innovations. https://smartcityhub.com/technology-innnovation/these-are-the-top-10-urban-innovations/. Accessible on July 2019
6. Dakin A, Burgess K, Adamson D (2012) CREW discussion paper 1: targeting area-based regeneration: methods and issues. Available at http://regenwales.org/resource_5_CREW-Discussion-Paper-1--Targeting-Area-based-Regeneration---Methods-and-Issues
7. Davoudi S (2006) Territorial cohesion, European social model and spatial policy research. In: Faludi A (ed) Territorial cohesion and European model of society. Cambridge, Mass, The Lincoln Institute for Land Policy
8. Dezhina (2011) Policy framework for innovation in Russia. EBRD, Mimeo
9. Elisei P, Leopa S, Drăghia M (2018) Territorial attractiveness monitoring platform. A handbook for policy planners, Bucharest. Available at http://www.interreg-danube.eu/uploads/media/approved_project_output/0001/25/064d81849140676a3f2a51001d860b3d710abbd3.pdf
10. Elisei P (2014) Strategic Territorial Agendas for small and middle sized towns and urban systems. https://issuu.com/sabinadimitriu/docs/status-book_final_web
11. ERA.Net RUS Plus https://www.eranet-rus.eu/index.php. Accessible on July 2019
12. Hall P (1998) Cities in civilization
13. Harfst J (2006) A Practitioner's Guide to Area-Based Development Programming, UNDP Regional Bureau for Europe & CIS. Available at https://www.undp.org/content/dam/undp/library/corporate/speakercorner/a-practitioner-guide-to-area-based-development-programming.pdf
14. Healey P (1997) Collaborative planning: shaping places in fragmented societies. Macmillan, Houndsmills, England. Available at https://books.google.pl/books?id=xkxdDwAAQBAJ&lpg=PR10&ots=qS1K28MKIg&dq=HealeyCP.(1997).CollaborativeplanningAShapingplacesinfragmentedsocieties.HoundmillsEnglandAMacmillan&lr&hl=ro&pg=PA297#v=onepage&q=challenge&f=false
15. Investment passport of Rostov-on-Don (2018) Rostov-on-Don city administration. Available at http://investrostov.ru/pages/investytsyonnyi-pasport-goroda
16. Koshkin P (2015) Turning Russia into an innovative country is still a challenge
17. Meyer PB 1998 Accounting for differential neighborhood economic development impacts in site-specific or area-based approaches to urban brownfield regeneration. Working paper. Center for Environmental Management, University of Louisville, Louisville, KY, 15

p. Available at https://www.researchgate.net/publication/251169893_Accounting_for_Differential_Neighborhood_Economic_Development_Impacts_in_Site-Specific_or_Area-Based_Approaches_to_Urban_Brownfield_Regeneration. Accessed 28 May 2019

18. OECD (2011) Review of Innovation Policy: Russian Federation, Paris
19. Science for Environment Policy: Brownfield Regeneration (2013) Thematic Issue 39. European Commission's Directorate-General Environment, Science Communication Unit—University of the West of England (UWE) Bristol
20. Stanilov K (ed) (2007) The post-socialist city: urban form and space transformations in Central and Eastern Europe after socialism. Springer, The Netherlands
21. Stelzlea B, Noenniga Jörg R (2017) A database for participation methods in urban development. In: International conference on Knowledge Based and Intelligent Information and Engineering Systems, Marseille, France, Elsevier B. V. Available at https://www.researchgate.net/profile/Benjamin_Stelzle/publication/319445455_ADatabase_for_Participation_Methods_in_Urban_Development/links/5a90159645851535bcd476f1/A-Database-for-Participation-Methods-in-Urban-Development.pdf
22. State of the Innovation Union (2015) https://publications.europa.eu/en/publication-detail/-/publication/0487b7b9-b5d6-11e5-8d3c-01aa75ed71a1/language-en/format-PDF. Accessible on July 2019
23. Tosics I (2019) New urban planning: long-lasting innovation or just a temporary illusion? https://www.blog.urbact.eu/2019/01/new-urban-planning-part-1/. Accessible on July 2019
24. UN, World Urbanization Prospects: The 2018 Revision (2017) Insider's guide to Russian high-tech hubs. http://www.russia-direct.org/catalog/archive/russia-direct-report-insiders-guide-russian-high-tech-hubs
25. Urban world: Mapping the economic power of cities (2011) https://www.mckinsey.com/~/media/McKinsey/Featured%20Insights/Urbanization/Urban%20world/MGI_urban_world_mapping_economic_power_of_cities_full_report.ashx. Accessible on July 2019
26. World Bank (2011) Reliving the Sputnik moment: innovation strategies for the post-transition economies. Washington, D.C.

From Information-Communication City to Human-Focused City

Pierre Laconte

Abstract Smart cities are using information and communications technologies—ICTs—to connect urban activities hitherto unconnected. ICTs can help connect activities within buildings, neighbourhoods or cities and help linking such activities as land use, heritage conservation, energy savings, telecommunications, commerce/banking and mobility. This functional concept can serve multiple aims and objectives, among others the production of knowledge-based services making use of "big data" collecting. Cities can appeal to their citizens and visitors by their quality of life. Beyond the gross development product statistics, quality of life includes perceived quality of air, water and health. The continuity of their urban landscapes invites to leisure activities ("green and blue" trails). It offers diversity of visual experiences by users of the public spaces ("views from the street" rather than "views from the road"). It offers squares, trees and gardens, fountains and canopies, all designed for both walking and sitting users, clean air and an overall urban density propitious to informal contacts between persons and generations. It encourages non-motorised mobility, for the sake of health and fossil fuel energy saving. That enhances availability of urban services, safety and security for all citizens, and on the other hand, the quest for "human well-being", a qualitative appeal to their citizens and users. This is perhaps a key to urban sustainability and adaptation to unavoidable economic, social and disruptive technical changes affecting cities.

Keywords Sustainability · Functional · Data · Quality of life · Urban services · Human · Emotional · Platforms · Circularity

1 "Smart Cities" as Functional Concept

"Smart cities" are using information and communications technologies—ICTs—to connect urban functions and activities hitherto unconnected.

P. Laconte (✉)
Foundation for the Urban Environment, Kortenberg, Belgium
e-mail: pierre.laconte@ffue.org
URL: https://www.ffue.org

© Springer Nature Switzerland AG 2020   27
V. Popovich et al. (eds.), *Information Fusion and Intelligent Geographic Information Systems*, Advances in Geographic Information Science,
https://doi.org/10.1007/978-3-030-31608-2_3

Information and communication technology can help connect activities within buildings, neighbourhoods or cities such as land use, heritage conservation, energy savings, telecommunications, commerce/banking and mobility.

This functional concept can serve multiple objectives and business interests, among others, the production of knowledge-based services making use of "big data" collecting. This will be illustrated by five examples.

1.1 Using Individual Data for Money Transfers: Mobile Phones as Banks

A kiosk displays M-Pesa advertising in a slum area of Nairobi, Kenya (2012) (Fig. 1). M-Pesa (Mformobile, Pesa is Swahili formoney) is amobile phone-basedmoney transfer, financing and microfinancing service, launched in 2007 by Vodafone in Kenya and Tanzania. The service allows users to deposit money into an account stored on their own cell phones, to send balances using PIN-secured SMS text messages to other users, including sellers of goods and services, and to redeem deposits for regular cash money.

M-Pesa is thus a "smart" branchless banking service.

Fig. 1 Kiosk displaying M-Pesa advertising

M-Pesa has spread quickly and by 2010 had become the most successful mobile phone-based financial service in the developing world. An estimated total of 17 million M-Pesa accounts have been opened in Kenya, and the service has expanded to Tanzania, Ghana, Afghanistan, South Africa, India and Eastern Europe.

The service has been lauded for giving millions of people access to the formal financial system and thus for reducing urban crime for cash in largely cash-based societies. This case was referred at in Bisello [3, page 4].

1.2   Big Data Collective Exchange Platforms Between Users and Between Users and Rulers

Data exchanges have been multiplying the traditional merchant exchange added value by getting to know the profile of the buyers. Big data collection generates value-added information of electronic platforms in fields such as accommodation (Airbnb), mobility (Uber), auctions (eBay) and collection of personal data (Facebook). The profiles so collected are a highly saleable product. This high value creation gives them easy access to international funding ("uberisation" of services). The data made available on each citizen also multiply the potential of citizen alienation by the masters of Big data collection, including governments (Fig. 2).

Beyond Huxley's "Brave New World" and Orwell's "1984", the Government of China is using big data for imposing its "Conformity policy" to all citizens. It is an "unbridled" toy for "unbridled" governments.

More modestly, smart platforms can be owned by independent cooperatives, a modern extension of the nineteenth-century cooperative movement (initiated by Manchester Rochdale cooperative pioneers). These new cooperatives are reported in the "Platform Co-operativism Consortium"—PCC (Scholz [10]).

Their profits in that case flow back to their members, instead of shareholders.

1.3   Smart Buildings as Power Plants and Resource Savers

Smart buildings aim at optimising air, water, energy production (solar energy generated from both roofs and windows), batteries charging, etc., within a centralised system. A "Smart Building" integrates major building systems on a common network and shares information and functionality between systems to improve energy efficiency, operational effectiveness, and occupant satisfaction.

The multinational contractor BESIX [2] sees its Dutch headquarter "smart building" as an encyclopaedia of smart building features (Fig. 3).

Fig. 2 A postcard showing the interior of Stateville Correctional Centre, Illinois, modelled on Bentham's Panopticon [1, 5]

Fig. 3 View of BESIX new Netherlands Headquarters, Dordrecht

Fig. 4 Small van fleets in service in Lyon (Photo Keolis, 2017)

1.4 Smart Mobility, Tool for Management of Autonomous Connected Vehicles

Autonomous vehicles (AVs), or zero-occupancy vehicles, have been described as a liberation from the driving chores.

However, according to the International Transport Forum [4] report, trucks without drivers should increase road congestion.

The reduction of the employment costs may indeed encourage fleet owners to put more—half empty—trucks on highways.

As to cars, AVs may in effect result in users acquiring greater tolerance for long-distance commuting and therefore increase urban spread, unless collective transport AV small van fleets are in service for short-distance links to access mass public transport, using the street network (Keolis 2017) (Fig. 4).

1.5 Saving Resources through Circularity

The principle is to replace the linear production chain (produce, use and throw away) by a circular production chain (produce, use and reuse into a secondary product). This can apply to many types of goods, including entire buildings.

A recent example of building recycling is the Amsterdam CIRCL pavilion (Fig. 5).

Fig. 5 "CIRCL", Amsterdam multi-purpose circular pavilion (ABN AMRO Bank 2017)

Every resource used in the building can be recycled. All parts of the building are dismountable.

One may remember that the Eiffel Tower was to be dismantled after the Paris World.

Exhibition was however kept ever since, by popular demand (Fig. 6).

Recycling in a wider context is only made possible by linking supply and demand for secondary products and certifying them. This may include recycling of CO_2 emissions instead of trying to store them underground.

2   "Human-Focused" Cities

Cities may well combine on the one hand the "smart cities" functionalities that enhance availability of urban services, safety and security and, on the other hand, the "emotional" qualitative appeal to their citizens and users.

2.1   Emphasis on Quality of Life, Leisure Activities and Education

"Medellin Ciudad Inteligente" encourages popular IT education, including a network of large and small libraries, even in metro stations (Fig. 7).

Fig. 6 Inside view of the same building

Fig. 7 España Library of Medellin. A strong statement in favour of "knowledge city". *Photo* by Municipality of Medellín. (Source: https://healthymedellin.weebly.com)

Fig. 8 New York City Times Square (before and after pedestrianisation). Photo source: https://voyagesmicheline.com

2.2   Enhancement of Citizen Satisfaction through Making Places Available for Informal Contacts

New York's Times Square was ever clogged by traffic. Mayor Bloomberg made pedestrianisation acceptable by introducing a trial period, during which traffic flows were analysed. This analysis showed it took less time for taxis to take alternative routes.

This was a smart combination of smart data and enhancement of citizen satisfaction (Figs. 8 and 9).

Rome's Piazza Navona may be seen as a human-focused urban place. It is not only an attraction but also as a form of urban theatre, as the God Neptune of Bernini's fountain is spewing towards the facade of his rival Borromini's Sant Agnese Church (Fig. 10).

2.3   Emotional Citizen Involvement through Community Events (Festivals, Cultural Events, Folklore, Supporter Sports)

See Figs. 11 and 12.

Fig. 9 Same view after pedestrianisation. Photo © New York Department of Transportation

Fig. 10 Rome's Piazza Navona. *Source* Port of Rome, Portale turistico del Porto di Roma (Source: http://www.portofrome.it)

Fig. 11 Oktoberfest, Munich, is the top yearly event of this 1.5 M inhabitant's city. Source: muenchen.de, Das offizielle Stadtportal (https://www.muenchen.de)

Fig. 12 "Carnaval des Gilles de Binche"—on UNESCO's World Heritage List—involves the entire population of Binche, a Belgian town of 30,000 inhabitants. Source: *Le Soir*, 28/02/2017. Photo © EPA

2.4 Human Focus in a Changing Climate Context

At a planetary level, greenhouse gases (GHGs) emissions are affecting climate. In cities, the focus could rather be put on the role of reduced fossil fuel energy pollutions affecting human health as a quality of life indicator. In other words, is a GHG reduction a correct indicator of citizen-focused quality of life in a climate change context?

A survey of GHG-emissions accounting methods has been done in a comparative study by Baader and Bleidschwitz (2009). While the measurement of GNP is made by public agencies according to internationally agreed methods, there is no such agreement about the six GHGs identified by the IPCC.

Who is entrusted with measuring?

GHGs measurement is done by independent institutions, e.g.:

1. CO_2 Grobbilanz/EMSIG (Climate Alliance Austria, Energy Agency of the Regions)
2. ECO_2 Region (Climate Alliance, Ecospeed)
3. GRIP (Tyndall Centre, UK Environment Agency)
4. Bilan-Carbone (ADEME, France)
5. CO_2 Calculator (Danish Environmental Research Institute)
6. Project 2 Degrees (ICLEI, Clinton Climate Initiative, Microsoft).

What emissions are measured and how?

1. The measurement covers either all of the IPCC GHGs or only some of them, mainly carbon dioxide and methane.
2. Different potential global warming estimates are obtained according to whether the second, third or fourth IPCC report is used.
3. The reporting standards are different.
4. The scope of measurement either only includes direct emissions or also includes indirect and life cycle emissions.

Measuring energy production and consumption as an alternative to direct measuring of CO_2 emissions?

Considering the political impossibility of a world agreement on the calculation of CO_2 and other greenhouse gases, the second best could be to analyse in depth the origin, production and use of fossil fuels, which according to Nicholas Stern [11] constitute the largest of the GHG emissions.

3 Global Urban Sustainability Includes both Functional and Human-focused Features

Hereafter we have been selecting three award winning cities/neighbourhoods within the list of European practices referenced in the literature assessing the sustainability performance of cities. These three awards are referred at in the book "Sustainable cities—Assessing the Performance and Practice of urban Environments" [6] (Fig. 13).

3.1 Zurich

Zurich may be considered as a smart mix of land use and mobility policies aimed at improved quality of life (1985–).

Fig. 13 Title page of the book "Sustainable Cities - Assessing the Performance and Practice of urban Environments"

Zurich's traffic management

In Zurich, trams and buses enjoy absolute priority on-street. When approaching a traffic light, the sensor (seen on the lower left) ensures they have a green light at any time of the day. The reliability of timetables makes public transport the city's fastest mode of transport. Modal split is around 80% in favour of public transport.

The sensors (lower left) trigger the traffic light priority in favour or trams and buses, not taxis (Fig. 14).

The political ingenuity, however, lies in the parking policy favouring local voters: the KISS Principle ("Keep it Smart Simple").

Zurich's parking management

Unrestricted on-street parking is exclusively reserved for Zurich-registered residents (the voters), while cars entering the city from other municipalities have a max. 90' free parking time. This measure has triggered a large-scale return of inhabitants to the city, has benefited the off-street car parks and has been politically very rewarding for the city fathers, while suburban rail travel patronage has been improved. This system could be applied in any city where commuters come from other electoral districts (Fig. 15).

Emotional attachment to the city's way of life is embodied by its attitude to mobility and relation between centre and periphery, e.g. Limmat Valley [9]. The inclusionary approach to mobility is illustrated by the Nissan automobile poster (Fig. 16).

Fig. 14 Zurich's traffic management

3.2 Bilbao

Bilbao is a smart mix of urban regeneration and multimodal transport, putting emphasis on culture, quality of life and public meeting places (1989–2012).

The long-time prosperous steel industry was wiped out by the 1989 steel crisis. Industrial land was re-used for new activities, based on services and culture, while preserving architectural heritage (Fig. 17).

The derelict industrial area along the Ría was owned by several public bodies, from local to national. This ownership was unified by a public–public partnership embodied in a public redevelopment corporation: Ría 2000. The two planned anchors for new development, at each end of the site, were the new Guggenheim museum and the congress and concerts centre (Fig. 18).

The valuable land situated between the two anchors and very close to the central business district was redeveloped by Ría 2000, with an obligation to invest all of the proceeds in new public infrastructure along the same canal.

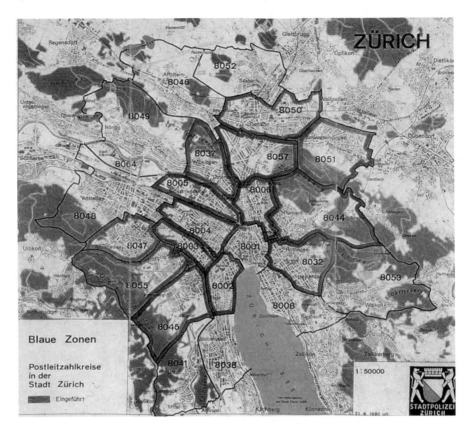

Fig. 15 City map showing the areas for unlimited parking by residents (90 minutes for non-residents). Source: Stadtpolizei Zurich

Fig. 16 Nissan automobile poster

Fig. 17 View of Bilbao's Ría industrial landscape prior to the 1989 steel crisis. Source: Archivo Municipal de Bilbao

The huge surplus generated by the land sales was to be used exclusively to enhance connectivity and further urban regeneration.

The plan's implementation was completed in 2011 by an office tower by Architect C. Pelli (Fig. 19).

A new tram line serves (Fig. 20) the canal side in the urban centre, saving traffic and parking space and adding to the citizens' health and quality of life.

Fig. 18 Cover page of Urban Land Europe, Autumn 2003 [7]

Fig. 19 Office tower. *Photo* P. Laconte

Fig. 20 New tram line in Bilbao

Bilbao Metro (Fig. 21) partly new (stations designed by Norman Foster) and partly reusing old industrial railways, it enhanced connectivity throughout the city and its region and attracted energy saving public transport.

Fig. 21 Bilbao metro

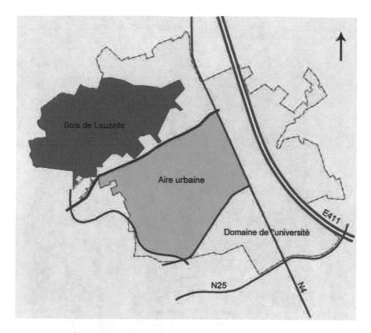

Fig. 22 Location of the new university town

3.3 *Louvain-la-Neuve (Brussels Conurbation)*

High density-low rise land-use—Planning for resource saving—is a key feature of the Louvain-la-Neuve new university town (conceived from 1968) [8] (Fig. 22).

The university bought 920 ha of agricultural and forest land in a rural area close to Brussels—Namur road (N4): the central part was set aside for urban development; forest land in the North was preserved. The overall master plan and architectural coordination—starting in 1968—was entrusted to the Groupe Urbanisme—architecture (R. Lemaire, J.-P. Blondel and P. Laconte) (Fig. 23).

The only connection point of the site to existing road infrastructure is the Brussels–Namur trunk road. Development started from there along an East–West pedestrian spine. Automobile access to buildings is ensured through side alleys (Epstein 2008).

Planning for uncertainty (stop and go)

As the project could be suspended at any time by political decisions development took place according to a linear pedestrian central spine, this allowed a step-by-step mixed urban development. Automobile access to buildings and parking are located outside of the spine, with occasional underpasses.

The pedestrian option allowed a reduction of heavy infrastructure costs and enhanced air quality and citizens' health (Fig. 24).

It was implemented from 1972 as the main street's first phase. It started from the existing road east of the site (right part of the picture), It was later extended to

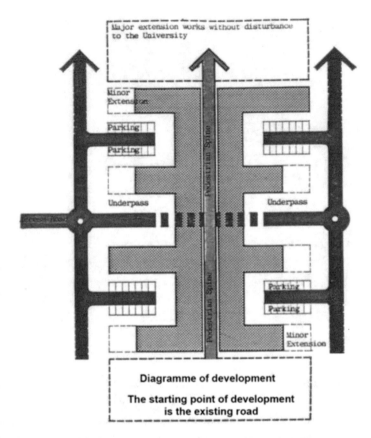

Fig. 23 Diagram of development of the central urban area of Louvain-la-Neuve

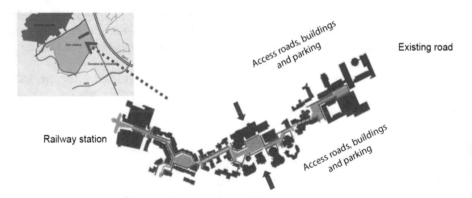

Fig. 24 Pedestrian place-making

Fig. 25 A string a public spaces alternating tranquility and animation. *Photo* P. Laconte and Koen Raeyemaekers

the railway station opened in 1975 (left part), the centre of the city, and its extension towards the western part of the site. Car access to buildings and parking takes place outside the spine, with some underpasses. Property development of the whole university-owned site (920 ha) is by long-term leases (75–99 years).

The centre of the first phase (1972) was the Science Library (Fig. 25), a huge concrete building seen as the cathedral of a university town with its plaza, above an automobile underpass. It is a social gathering place with university buildings, shops and restaurants (arch. A. Jacqmain). Built 45 years ago, it has consistently been a place alternating tranquillity and animation (Fig. 26).

All streets are pedestrian and combine university buildings, housing, retail and cultural services. Land remains the property of the University and is leased to investors. All motorised transport is located underground.

The diagram shows how the ground below—essential for long-term connectivity—remains the property of the university. The infrastructure and buildings are leased to public and private investors. High-density, low-rise development includes no high-rise buildings (Fig. 27).

View on one of the small piazza's forming a string of public spaces along the pedestrian street network. Trees are growing on the concrete slab. Cars are parked underneath (Fig. 28).

The shopping mall adjacent to the railway station (8 million visitors/year) and the private Hergé museum (arch. de Portzamparc, Paris) are all part of its high-density, low-rise development (Fig. 29).

All storm water is led to a reservoir which is treated as a lake, which saves infrastructure costs and attracts residential investment (Fig. 30).

Louvain-la-Neuve, example of human-focused smart development

The continuity of its planning—according to the initial master plan—and its governance, over the first 50 years of the up to 99 years leases, were ensured by the

Fig. 26 Street entrance to the railway station. *Photo* P. Laconte

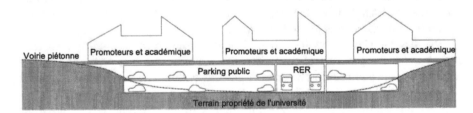

Fig. 27 Functioning of the slab. *Source* UCL Archives

combined strength of its mayors—the latest one over 18 years—and of its residents' council (Association des habitants de Louvain-la-Neuve), strong countervailing power to both the city authority and the university land lord.

Fig. 28 Numerous small piazzas. *Photo* P. Laconte

Fig. 29 Shopping development. *Photo* Hergé Museum

Fig. 30 Louvain-la-Neuve water management. *Photo* Wilco 2011

4 Conclusion

Combining, on the one hand, the "smart cities" functionalities—helped by GIS deployment—that enhance availability of urban services, safety and security for all citizens and, on the other hand, the "emotional" qualitative appeal to their citizens and users is perhaps a key to urban sustainability and adaptation to unavoidable economic, social and disruptive technical changes affecting cities.

Appendix

An artist illustration

Mobility and Liveable Cities—the transport network irrigating the city—poster by
Friedensreich Hundertwasser (1928–2000) for UITP (1991).

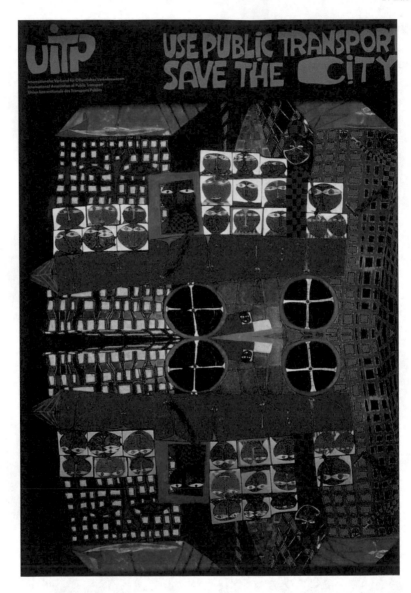

Mobility and Liveable Cities—the compact city—poster by Friedensreich
Hundertwasser for UITP (1993).

Mobility and Liveable Cities—enjoyment as a key to liveability—poster by
Friedensreich Hundertwasser for UITP (1995).

References

1. Bentham J (1791) Panopticon or the inspection house. London
2. Besix (2018) Besix NL HQ—100 years Excellence
3. Bisello A et al (eds) (2018) Smart and sustainable planning for cities and regions. Results of SSPCR 2017. In: Green energy and technology. Springer
4. International Transport Forum—ITF (2017) ITF Report: the future of autonomous trucks
5. Jaivin L (2014) The end of secrets. The Monthly, 6/2014
6. Laconte P, Gossop C (eds) (2016) Sustainable cities: assessing the performance and practice of urban environments. London/NY, IB Tauris
7. Laconte (2003) Smart growth in Spain, ULI Europe 2003
8. Laconte P (2016) The founding and development of Louvain-la-Neuve, the only new town in Belgium—in historical perspectives—history, urbanism, resilience, vol 5. In: International planning history society proceedings, 17th IPHS conference. Delft
9. Scholl B et al (ed) (2018) Spatial planning matters! ETH Zürich
10. Scholz T (2016) Platform cooperativism: challenging the corporate sharing economy. Rosa Luxemburg Stiftung, New York City. Print
11. Stern N (2018) New climate economic report. https://newclimateeconomy.report/ (latest reading 27.04.2019)

A Density-Based Measure of Port Seaside Space-Time Utilization

Behnam Nikparvar and Jean-Claude Thill

Abstract The space-time study of seaside space use by vessels may support traffic and operation management practices at ports. In this paper, we quantify space utilization using a volumetric index resulting from a density-based generalization of vessel movements. We use an extension of the kernel density estimation method in a 3D space where the third dimension is time and voxels are the basic elements of this space. Vessel trajectories are represented as line features passing through these voxels. To measure the utilization of space at different times, we aggregate the presence of vessels in a voxel using a prime shape kernel. We applied this approach to study space utilization at a terminal in the Port of Rotterdam during September 2017. Our results show that this measure of space-time utilization is practical and informative to monitor how intensively vessels use space across the waterside assets of the port, along the time line and at various temporal scales.

Keywords Space-time utilization · Space-time cube · Vessel trajectory · Port

1 Introduction

In concert with the fast-expanding globalization of the economy, the number of commercial vessels that call at ports has been increasing dramatically and the proper use of space within the confines of these points of contact between sea-based and land-based freight movements is important to manage the traffic in safe, reliable, and efficient ways. Seaside space utilization becomes even more important in ports that the topographic and geographic setting may limit the management of traffic and

B. Nikparvar (✉)
Infrastructure and Environmental Systems Program, University of North Carolina at Charlotte, 9201 University City Blvd., Charlotte, NC 28223, USA
e-mail: bnikparv@uncc.edu

J.-C. Thill
Department of Geography and Earth Sciences, University of North Carolina at Charlotte, 9201 University City Blvd., Charlotte, NC 28223, USA
e-mail: Jean-Claude.Thill@uncc.edu

© Springer Nature Switzerland AG 2020
V. Popovich et al. (eds.), *Information Fusion and Intelligent Geographic Information Systems*, Advances in Geographic Information Science,
https://doi.org/10.1007/978-3-030-31608-2_4

55

operations [1]. How do we know if port operators are using space efficiently and in a safe way? And how do we know if we are using a specific part of this space properly in different time periods in line with principles of maximization of return on investment?

The vessel trajectory data, which are available thanks to the automatic identification system (AIS), form one source that we can leverage to see how ships, historically, use the space of ports. To answer the above questions, we also need to choose between the two existing conceptual models to study traffic at ports: a network (topological) space or a Cartesian space. From a network point of view, we may define a measure of space utilization as attribute of a specific place (node) such as the number of vessels that docked at a specific berth or the capacity of the waterway (link) that connects this berth to an anchorage location [2]. This gives a simple measure and understanding of the traffic and space utilization at ports, and it may be available for different time periods. The downside is that it is not capable of representing the utilization of space in an explicit geographic space and especially in the case of maritime transportation where the movement is largely unimpeded in all directions. In such case, we are unable or have extreme difficulty to identify which parts of the waterway we use less intensively than the other parts. We may achieve this by dividing the links into smaller segments and measure the capacity of each one, but in case we need to investigate trajectories that cross a specified waterway or within an anchorage area, this approach is ineffective.

In a Cartesian space, the vessels are free to move in a 2D space and this is more likely to fulfill the goal and needs of the present work. In this framework, we may simply count the number of vessels recorded in each cell of a grid or estimate the density using available trajectory data for a specific geographic area of interest. We can delineate the approximate boundary of waterways [3] or optimize the arrangement of other traffic areas for a single or multiple time intervals [4]. This density-based model provides a sense of space utilization within ports. However, if the number of time periods increases, then this approach is not easy to work with in terms of visualization and decision analytics.

Bian suggested an object-oriented representation of fishes in an aquatic environment using a 3D Cartesian coordinate system where she modeled the movement of fishes with point features during a period of time [5]. In that case, the third dimension is depth and in order to represent the change over time she used several 3D snapshots. A space-time cube can be defined as a Cartesian framework where the third dimension is time [6]. We can extend the vector and raster data models to this space. This space is of special relevance to represent moving object trajectories. We may represent trajectories of moving objects in space and time as line segments and nodes. However, this may just provide a visualization of the space use, and even in that case, this is not practical when we have a large number of vessel trajectories [7]. For the purpose of this paper, we use the extension of the raster data model and of density estimation from a 2D space where it provides an aggregated value for traffic in elements of space and time. In this case, the elements of the space are turned into volumetric cells (voxels) where the third dimension is time.

This paper presents the principles of this approach and their implementation on actual data. We selected the Port of Rotterdam as the study area. The main reason to select this port is that it is one of the main hubs in global maritime transportation, where space use in terms of efficiency and safety is an acute concern. We use 15-min interval historical data of vessel movements from the automatic identification system (AIS) for the period of September 2017.

The rest of the paper is structured as follows. Section 2 reviews the main body of literature on space-time cube and on density estimation in this model. Section 3 presents the concept of space-time cube and the two main approaches that we selected to implement in this work. Section 4 describes the data and study area. Section 5 presents and discusses the results of our Port of Rotterdam use case. Finally, Sect. 6 presents the conclusions of this work and suggests future extensions of this spatial data fusion approach.

2   Literature Review

For trajectory data, the time dimension is important as much as the spatial dimension because it has a lot of variations that may cause complex patterns. A decent body of literature is available on the visualization of vessel movement, especially for large amounts of data. Andrienko and Andrienko used a vector-based space-time cube to visualize vessel movements [7]. They attempted to address the problem of cluttering in this approach by adding interactivity to visualization. The effectiveness of a simple 2D kernel density approach to visualize a large volume of movement data has been studied in the literature, with a conclusion that it has great merit [8]. In order to see the overall behavior of the vessels in different time periods, one approach is to use the convolution of a kernel and trajectories for different time stamps and visualize how the density or intensity of the moving objects changes using multiple small maps or animations. By generating kernel density maps for different time intervals during a day, [9] added these maps together in order to keep the details of all periods. This helps to show the details of diverse patterns exhibited over the day, but the problem is we still do not know which pattern in density belongs to which time period and their longitudinal sequence. Adrianko et al. [10] introduced the concept of time masks to interactively visualize large movement datasets where time intervals that fulfill some conditions are used as a filter in query to reduce processing time and memory allocation. Besides, visualization is a common way to recognize change in patterns that represent the use of space, but it is not the end goal of this work.

Monitoring and decision-making systems are an area where analysis of trajectory data may contribute to the processes involved in maritime logistics and transportation [11]. Chen et al. quantify the approximate boundary of principal fairways (main waterways) using a line-based kernel density estimation in 2D space [3]. They borrow an approach from animal space use pattern studies; specifically, they use the concepts of utilization distribution, home range, and core area to qualitatively determine boundaries of principal fairways. They followed this approach to study changes in boundaries over different seasons. However, for finer temporal resolutions, this approach is still not practical.

Desjardins et al. used the space-time cube to represent the pollen counts for elements of space and time where the centroid of these elements is the zip codes [12]. They computed the count of pollen instances within a neighborhood of each element of space and time after interpolation. In case of moving objects, we can represent the trajectory of objects as consecutive points in space and time and count their number within each voxel. This time we can represent changes in time properly but the downside is that we underestimate the space use between consecutive points. Demšar and Virrantaus used the trajectories as line features and computed the density of trajectories in a neighborhood of a voxel to address this problem [13]. We use both approaches to compute the frequency and density of the vessels within a voxel. We expect this view of the space-time cube to provide a proper generalized representation of the space utilization by vessels in ports.

3 Space-Time Cube

We represent space-time with volumetric cells (voxels) where the first two dimensions show the geographic elements and the third dimension represents time. A set of these voxels that abstract part of the geographic space and time line creates a space-time cube $C = \{v_1 = (x_{c1}, y_{c1}, t_{c1}), v_2 = (x_{c2}, y_{c2}, t_{c2}), ..., v_{cs} = (x_{cs}, y_{cs}, t_{cs})\}$, where x_c, $y_c \in R$ and $t_c \in R^+$ and s is the number of voxels in the cube. Each voxel may have different attributes. The main application of a space-time cube is either to generalize the occurrence of events (like the number of times a voxel is traversed by ships) or to represent aggregated values associated with a phenomenon or object in a voxel. Next, we need to conceptualize the trajectories. We represent the trajectory for object j by a set of consecutive points in space and time such as $T_j = \{p_1 = (x_1, y_1, t_1), p_2 = (x_2, y_2, t_2),..., p_n = (x_n, y_n, t_n)\}$ where $x, y \in R$ and $t \in R^+$ and n is the number of points in a trajectory.

For the purpose of this paper, we use the count and presence weighted by the normalized distance between a voxel centroid and the trajectory point to provide an absolute measure of utilization in each voxel. We do this either by counting the number of trajectory points that exist within each voxel or by assigning a value between 0 and 1 that is defined by the normalized distance from the trajectory point to the center of the voxel. In terms of computation time and complexity, representing trajectories of objects as point features is cost-effective. Most prior studies related to the space-time cube adhere to this approach (Fig. 1a). However, since movement is a continuous phenomenon and the sampling rate of observations is usually low, we have an underestimation of space utilization. Representing trajectories as nodes connected by line segments may improve the estimation (Fig. 1b) although it is highly dependent on the rate of sampling during observation and the method that

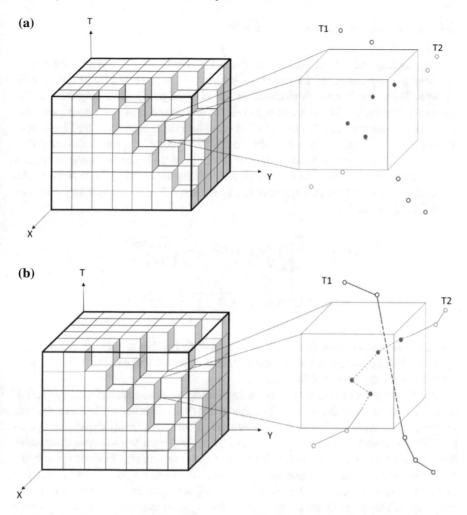

Fig. 1 Space-time cube. Trajectory $T1$ does not contribute to the density in point-based approach (**a**). Both trajectories contribute to density estimation in line-based approach (**b**)

we use for representing the segment between two consecutive nodes. In Fig. 1, $T2$ is involved in density estimation for both cases of point- and line-based methods. However, in case of point-based space-time cube (Fig. 1a), $T1$ has no contribution in density estimation for the voxel because there is no point instance of trajectory within the voxel volume. When we represent the trajectory as a line feature (Fig. 1b), we involve $T2$ to estimate density, which is expected to be more accurate. We explain these two methods in detail in the following sections.

3.1 Point-Based Space-Time Cube

In the simplest case, we count the number of points that exist within each voxel. We have the trajectories of the vessels as points with latitude and longitude as well as time. Thus, that is enough to count the number of points from trajectories that are located in each voxel. We can also weight the presence of each point with the normalized distance from the center of the voxel as a value between 0 and 1 and sum all those values together for a specific voxel. This second approach differentiates between points that are on the boundary of a voxel and those that are on the centroid of the voxel, or somewhere in-between by weighting them using the inverse of the associated distance. Thus, the space utilization for voxel v_i is computed by either of the following formulas:

$$\text{density}_{v_i} = \sum_{j=i}^{m} \sum_{k=1}^{n} c_k; \text{ where } c_k = \begin{cases} 1 \, k \text{ inside } v_i \\ 0 \text{ otherwise} \end{cases} \tag{1}$$

$$\text{weighted\_density}_{v_i} = \sum_{j=1}^{m} \sum_{k=1}^{n} \left(1 - \frac{d_{ik}}{r}\right) \tag{2}$$

where m represents the number of trajectories, n is the number of points in a trajectory, d_{ik} is the distance between voxel i and trajectory point k, and r is the distance from the centroid to the corner of the voxel.

One issue that is of notice when working with space-time density is the limitation related to the computation cost. In the case where distance is considered as a weight, we need to compute the distance in 3D which is a computationally expensive operation. To decrease the implementation time, we use an approach to avoid computing the distance to the trajectory points for voxels with zero density. For each trajectory point, we create a buffer around that point with the radius of the kernel (which in our case consist of the dimensions of the voxel) and generate the voxels only for these areas. This means that, for weighted_density$_{vi}$, we only compute the value for the points that are within the voxel. This approach reduces the computation time significantly because a huge area of the space-time cube is occupied by land and there is no trajectory point in that space. The downside of using point instances of a trajectory to estimate utilization is the underestimation results from the gaps between the point instances of a trajectory due to the temporal resolution of the data points. Using lines as opposed to points to estimate the density may improve the results.

3.2 Line-Based Space-Time Cube

In this approach, we consider the trajectory as a set of point features again. However, this time we connect those points with line segments and compute the density for each voxel according to these linear segments. We use the method suggested in [13].

The basic concept is a 3D kernel that moves around the line segments of trajectories and computes the count of trajectories that are within a neighborhood of each voxel. For each voxel, we compute either the count of trajectories or the distance from the centroid of the voxel. In the second case, for each trajectory that passes through a voxel, we compute the distance from the center of the voxel to all segments of that trajectory and then consider the minimum distance as the distance to the trajectory. In order to compute this distance, we consider the relative position of the center of the voxel to each line segment. The distance to a segment is the minimum distance to one of the end nodes of the segment or the perpendicular distance to the line. After computing the distance of a line to the voxel, it is used to weight the density. If the line is not within the voxel, then the density of the voxel for that line is 0. When the line is on the voxel, the density is 1. For any line within the kernel radius (which is equal to dimensions of a voxel in our work), the contribution to the density of the voxel is computed using a linear function that maps the density that is weighted inversely by distance between 0 and 1 (Eq. 2). This process is repeated for all trajectories, and the density of each voxel is computed based on the sum of intensity contributions of all trajectories. This value represents an absolute measure of space utilization at each voxel. Thus, we can develop the formula that we had for the point-based estimation but this time we count the presence of trajectory lines and their distance to the centroid of the voxel.

There is a challenge with this approach that is similar to that encountered in the previous section that makes the computation expensive in terms of time. The problem is that we are considering a neighborhood around each line to compute the density and weighting the density based on the distance of the lines to the center of a voxel. We need to compute distances, which is a time-consuming task when we need to repeat it for all segments of each trajectory and for all of the voxels in the cube. However, there are many voxels in the cube with zero density and we do not need to compute the distance to trajectory segments for them. To decrease the number of voxels that need processing, we can consider a bounding box around each trajectory according to the radius of the kernel and create the voxels that are within this bounding box [13]. This process is repeated for segments, and all trajectories and distinct voxels will be added to a single list. This way a large number of 0 value voxels will not be considered in the process of computing distance.

The other challenge here is the way that we define the line segments between consecutive points. In the simplest case, we can connect each pair of points using a straight line. However, if there is a physical barrier between two points, there is an intersection with that object when we use a simple straight line. One way to solve this issue is to use an approach that can estimate the line segment between two points so that it does not intersect the barrier and uses the shortest path to connect the two points together. There are still challenges with this approach since we need to consider environmental and object properties to be able to estimate a route that simulates the reality more accurately, but for the scope of this paper we focus on finding the shortest path between two consecutive points. We use the well-known A* algorithm to impute new points for the trajectory [14].

4 Data and Study Area in the Use Case

We selected a specific area in the Port of Rotterdam, the Netherlands, by considering a number of facts about this port. First, Rotterdam is one of the largest container ports in Europe, and it is considered as one of the key hubs in global trade. The other key reason that we selected this port is the importance of the space utilization because of the geographical constraints. Figure 2 represents the Port of Rotterdam and the specific terminal and waterway that we selected for this study. There is no specific consideration to select this part of the port other than being an area that includes both waterway and berth spaces. We separated berth and waterway because in berth areas the utilization/un-utilization and density of usage are both important but for the waterway the density is mainly of interest due to the risk of collision and crash. We treat a water area as a berth when the distance to the dock where ships can be moored is less than 150 m. The study area includes terminals that have recently been developed in the port through land reclamation.

We used AIS data that represent 15 min interval points for 897 trajectories of various types of vessels in the port area during the month of September 2017. Utilization estimation implemented in Python 2.7 and ArcGIS 10.6 products as well as Voxler 4.1.5 was used to visualize the results. Coordinates have been projected using Web Mercator Auxiliary Sphere (x and y in meters).

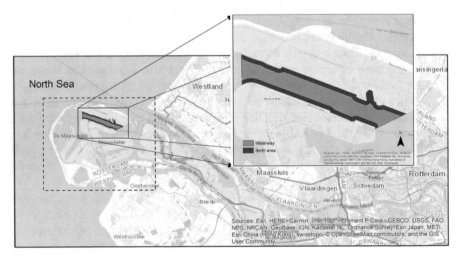

Fig. 2 Study area, Port of Rotterdam

5 Implementation and Results

We are interested in demonstrating how the space utilization measure can assist port management in answering two sets of questions. First, in the berth areas, we look for zones where space utilization is extreme, either patently underutilized (a sign of weak return on investment) or overutilized (where safety breaches may occur more readily). Second, in the waterways, we seek to identify areas of high vessel utilization, which can heighten the safety risks.

We start with the approach that we used to generate new instances between each consecutive pair of points in a trajectory to avoid possible intersection with terrain. We need to conceptualize the space as a grid to be able to use the A* algorithm. We extracted all existing trajectories for a larger area in the port (the red dashed line box in Fig. 2). This was to insure we simulate new instances for the pairs of consecutive points that are inside and outside of the study area so that we avoid intersection with the terrain. For this area, we created a grid so that pixels with center within the water side have value of 0 and pixels with center in the land side have value of 1. We used a dimension of 150 m * 150 m for this grid. The criterion we used to determine this value was that it is large enough to avoid high computation time while generating enough points to avoid intersecting the land side. Then, we used the A* algorithm to compute the shortest path between each pair of consecutive points where the distance between these points is larger than 100 m. Thus, we have two trajectory datasets: the original (dataset 1) and the new dataset (dataset 2). We used dataset 1 for the point-based approach and dataset 2 for the line-based approach. Then, we created a cube of voxels with dimension 100 m * 100 m * 1 h in the study area and for the first week of September 2017.

The results of utilization estimation with the two approaches presented above are very different. Figure 3 shows the corresponding frequency of nonzero counts of vessels for voxels in the space-time cube. The difference indicates that the utilization may be highly underestimated in the point-based method. At the same time, we need to keep in mind that our estimation of space utilization does not necessarily align with reality, especially because we are imputing parts of the trajectories.

Figure 4 represents an overview of the results for the first day of the week. The vertical axis is the time of the day, and the grid represents the footprint of voxels.

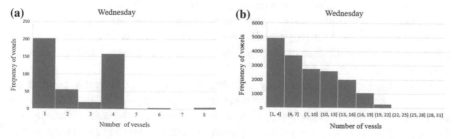

Fig. 3 Frequency of nonzero voxels. point-based approach (**a**); line-based approach (**b**)

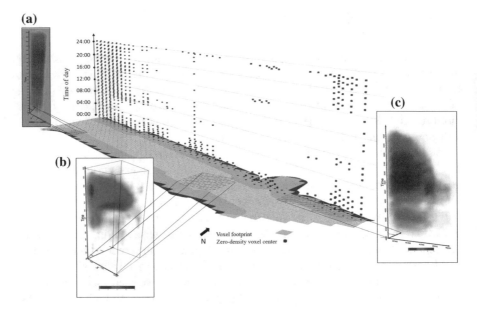

Fig. 4 Space utilization in berth and waterway

Small red cubes along the northern berth area represent the voxel centers with the value of zero. A zero value means that no vessel has passed the voxel over a one-hour period, so it is unutilized. We see the northwestern part of the terminal is almost unutilized (with a length of about 0.5 km along the berth). The berth was not used for the most part in the first four hours of the day. There are areas that have not been used during specific hours (see the right side of the birth). Sub-figures a, b, and c represent the density of vessel movements in two berth locations (a and c) and one waterway area (b). Utilization is lower in evenings and through the night. In all three locations, utilization reaches its peak in mid-to-late afternoon. This may suggest that some spread of vessel traffic from afternoon to others times may improve efficiency of the port infrastructure.

We can use the same cloud representation of the density to identify areas with higher rate of utilization across large pieces of the space-time cube (Fig. 5). We can clearly see the southern berth area has more traffic than the northern berth (a) and in two periods of time during the day (mid-morning and mid-afternoon), traffic is high for along most of this berth (c). Finally, we see that the density in the eastern part of the waterway is higher than in the west, and they may be areas with higher risk of collision or crash (b).

Figure 6 shows how the density is changing along specific spatial profiles (or transects) for a period of 24 h. We can see how a large portion of the berth during time is at low level of utilization (a). Representing density with cutting planes across different longitudes reveals interesting patterns of use (b). Moving from west to east, the utilization of southern berth is decreasing while the opposite is correct for the

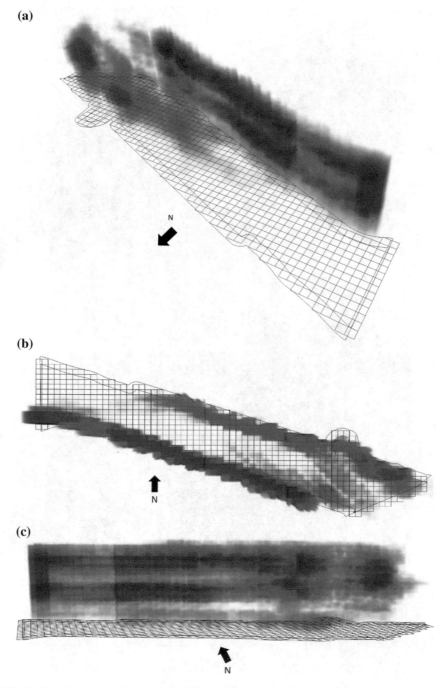

Fig. 5 Density cloud: **a** view from north; **b** view from top; **c** View from south

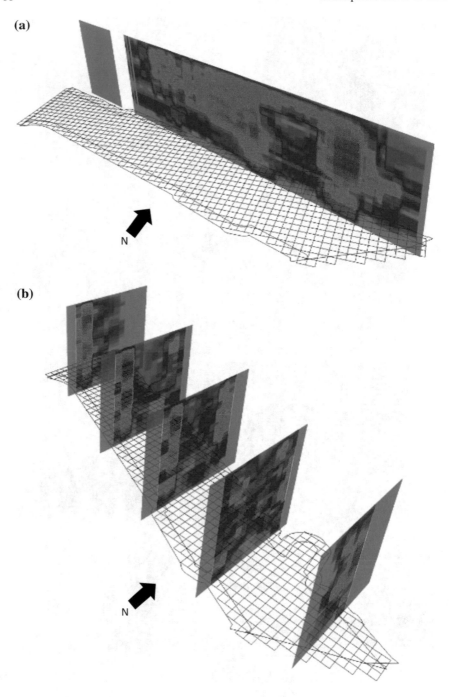

Fig. 6 a Utilization of space along a specific line; **b** utilization of space across several longitudes

northern berth area. Figure 7 is the same visualization over a one-week period. Daily repetition of space utilization can be discerned, with peaks often detected in the middle of the afternoon. Also, we readily see the area associated with the left slice has high traffic density during the week, and a hot spot is observable on Wednesday.

In addition to the visualization form presented so far to assist in pinpointing problem areas and times and in formulating solutions, results can also be reported in tabular form. Table 1 represents the frequency of voxels and the percentage of the space-time cube according to vessel count classes, separately for waterway and berth areas during the first week of September 2017. The shares of the entire space-time cube for both waterway and the berth area that are not being used are close, which means there is no specific difference between the two areas on the whole. However, we can see a significant difference between the two for the voxels with count value between 0 and 10. Part of it can be explained by the number of voxels that exist in each area. 54% of the voxels are in the berth and 46% are in the waterway area. However, this will reverse when we have voxels with value between 10 and 20 and the percentage of voxels in berth areas are much more than the waterway. This is meaningful because it is more probable for the large number of ships to be at berth in one hour and in a 100 m * 100 m area. There is no significant difference between days of the week in terms of the space that is not being used. Monday and Wednesday appear to be somewhat more congested in comparison with other days of the week, especially at berth areas. However, we would need to analyze other weeks as well to be able to make this inference more valid.

6 Conclusions

We used two approaches consistent with a space-time cube framework to estimate a measure of space utilization across the spatial and temporal dimensions, specifically for objects that are moving. All codes have been written in python to estimate density, while Arc Scene 10.6 and Voxler 4.1.5 were used to visualize the results. The comparison indicates that, in case of moving objects, the computationally simpler point-based approach is probably less reliable because of underestimation detected in our use case. Visualization of the space-time cube can provide valuable insights on how the port is being utilized, both in terms of overall distribution across the cube and in relation to specific parts of the cub that are more closely examined. Visualization can be challenging, however. To be more practical and useful for port space management, it needs to be more interactive and integrated with other tools to add value to the experience and knowledge. The point is we can select a specific berth and see how much the space-time cube is being used by vessels in that place. Some of the challenges we faced during the design and implementation of this work include (i) the way we impute new point values to trajectories to avoid intersection with terrain by considering environmental and object properties, (ii) computational cost, and (iii) proper visualization tools and platforms. These are all areas of interest for future research.

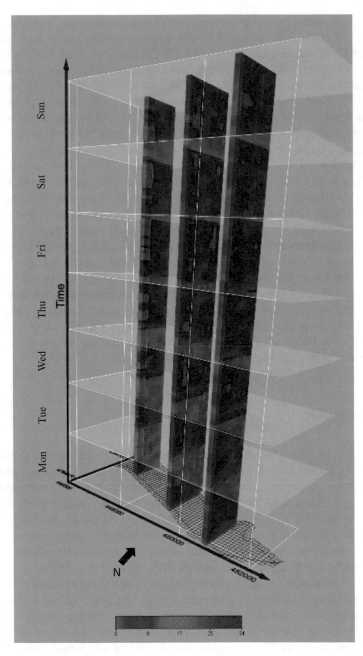

Fig. 7 Utilization of the space over a week for three longitudinal profiles

Table 1 Frequency of vessels in waterway and berth areas for 7 days of a week

Day (w = waterway, b = berth)	Total number of voxels	Count = 0		0 < count ≤ 10		10 < count ≤ 20		20 < count		Total
		Count	%	Count	%	Count	%	Count	%	%
Mon (w)	18,650	1961	10.5	7942	42.6	196	1.1	1	0.0	54.2
Mon (b)		1584	8.5	4695	25.2	2122	11.4	149	0.8	45.8
Tue (w)	18,650	1265	6.8	8781	47.1	54	0.3	0	0.0	54.2
Tue (b)		833	4.5	6374	34.2	1342	7.2	1	0.0	45.8
Wed (w)	18,650	1644	8.8	8283	44.4	166	0.9	7	0.0	54.2
Wed (b)		1264	6.8	5265	28.2	1744	9.4	277	1.5	45.8
Thu (w)	18,650	1335	7.2	8628	46.3	137	0.7	0	0.0	54.2
Thu (b)		1096	5.9	5791	31.1	1603	8.6	60	0.3	45.8
Fri (w)	18,650	1418	7.6	8514	45.7	165	0.9	3	0.0	54.2
Fri (b)		1213	6.5	5131	27.5	2152	11.5	54	0.3	45.8
Sat (w)	18,650	1544	8.3	8454	45.3	102	0.6	0	0.0	54.2
Sat (b)		1139	6.1	5781	31.0	1561	8.4	69	0.4	45.8
Sun (w)	17,158	1910	11.1	7223	42.1	159	0.9	0	0.0	54.2
Sun (b)		1270	7.4	4729	27.6	1812	10.6	55	0.3	45.8

References

1. Huang SY, Hsu WJ, He Y (2011) Assessing capacity and improving utilization of anchorages. Trans Res Part E Logistics Trans Rev 47(2):216–227
2. Fan HS, Cao JM (2000) Sea space capacity and operation strategy analysis system. Trans Planning Technol 24(1):49–63
3. Chen J, Lu F, Peng G (2015) A quantitative approach for delineating principal fairways of ship passages through a strait. Ocean Eng 103:188–197
4. Chen J, Lu F, Li M, Huang P, Liu X, Mei Q (2016) Optimization on arrangement of precaution areas serving for ships' routing in the taiwan strait based on massive AIS Data. In: International conference on data mining and big data. Springer International Publishing, pp 123–133
5. Bian L (2000) Object-oriented representation for modelling mobile objects in an aquatic environment. Int J Geogr Inf Sci 14(7):603–623
6. Kveladze I, Kraak MJ, van Elzakker CP (2013) A methodological framework for researching the usability of the space-time cube. Cartographic J 50(3):201–210
7. Andrienko N, Andrienko G (2013) Visual analytics of movement: An overview of methods, tools and procedures. Info Vis 12(1):3–24
8. Willems N, Van De Wetering H, Van Wijk JJ (2009) Visualization of vessel movements. In: Computer graphics forum, vol 28, no 3. Blackwell Publishing Ltd, pp 959–966
9. Scheepens R, Willems N, van de Wetering H, Van Wijk JJ (2011) Interactive visualization of multivariate trajectory data with density maps. In: Visualization symposium (PacificVis), 2011 IEEE pacific. IEEE, pp 147–154
10. Andrienko N, Andrienko G, Camossi E, Claramunt C, Garcia JMC, Fuchs G, Hadzagic M, Jousselme AL, Ray C, Scarlatti D, Vouros G (2017) Visual exploration of movement and event data with interactive time masks. Visual Info 1(1):25–39
11. Claramunt C, Devogele T, Fournier S, Noyon V, Petit M, Ray C (2007) Maritime GIS: from monitoring to simulation systems. Information fusion and geographic information systems. Springer, Berlin, Heidelberg, pp 34–44
12. Desjardins MR, Hohl A, Griffith A, Delmelle E (2018) A space–time parallel framework for fine-scale visualization of pollen levels across the Eastern United States. Cartography Geographic Info Sci 1–13
13. Demšar U, Virrantaus K (2010) Space–time density of trajectories: exploring spatio-temporal patterns in movement data. Int J Geogr Inf Sci 24(10):1527–1542
14. Hart PE, Nilsson NJ, Raphael B (1968) A formal basis for the heuristic determination of minimum cost paths. IEEE Trans Syst Sci Cybern 4(2):100–107

IGIS Integration with Acoustics and Remote Sensing

Modeling of Surveillance Zones for Active Sonars with Account of Angular Dependence of Target Strength and with Use of Geographic Information Systems

Vladimir Malyj and Andrey Mikhalchuk

Abstract In this paper, the features of an efficiency evaluation of active sonars on the basis of modeling and visualization of the expected surveillance zones with account of angular dependence of target strength moving in a given course are considered. The impact assessment of quality of information support of geographic information systems on the accuracy of calculation of the expected surveillance zones of active sonars in various hydrological and acoustic conditions is carried out.

Keywords Active sonar systems · Surveillance zone · Target strength · Dependence of target strength on the irradiation course angle · Hydrological and acoustic conditions · Inhomogeneous marine medium · Geographic information systems

1 Introduction

The surveillance zone (SZ) of the active sonar system (ASS) is constructed relative to its carrier and determines the spatial area at which the sonar target can be detected with a given probability of correct detection (PCD). In this paper, the modeling of an SZ ASS is considered in relation to the typical task of searching for an underwater target solved by a surface ship (SS) using the sonar operating in active mode.

The SZ ASS models are considered both for theoretical conditions of an infinite homogeneous medium (IHM) and for more close to real ones—for conditions of a layered inhomogeneous medium (LIM) with borders in the form of a flat bottom and surface. At the same time, in traditional models of SZ ASS, there is usually a range of ASS calculated for a fixed sonar target visibility (target strength (TS)) and do not take into account the dependence of the TS on the irradiation course angle (CA),

V. Malyj (✉) · A. Mikhalchuk
SPIIRAS Hi Tech Research and Development Office Ltd, 39, 14 line, Saint Petersburg 199178, Russia
e-mail: maliy@oogis.ru

A. Mikhalchuk
e-mail: mikhalchuk@oogis.ru

© Springer Nature Switzerland AG 2020
V. Popovich et al. (eds.), *Information Fusion and Intelligent Geographic Information Systems*, Advances in Geographic Information Science,
https://doi.org/10.1007/978-3-030-31608-2_5

which leads to a significant simplification of the models under consideration and, accordingly, to significant errors in estimating real SZ.

As the analysis showed, in order to increase the accuracy of the estimates of SZ ASS, it is necessary to take into account the dependence of the TS on the irradiation CA, and this problem can only be solved using modern geographic information systems (GIS).

2   Surveillance Zone Model of the Surface Ship Sonar System Without Taking into Account the Dependence of the Sonar Range on the Target Strength and the Course Angle of Its Irradiation (for Conditions of Infinite Homogeneous Medium)

The results of the modeling a typical modern SS ASS on a target with fixed sonar visibility (equivalent sphere radius $R_e = 10$ m, $TS = 20\lg(R_e/2) = 13.98$ dB) for conditions of the IHM are shown in Figs. 1 and 2.

As a rule, it is customary to consider the limit of SZ ASS for a fixed value of the PCD, equal to $P_{cd} = 0.9$. In this case, the border of SZ, taking into account the real surveillance sector, is a circle with a radius equal to the sonar range in IHM D_0 with

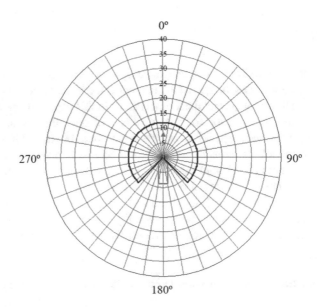

Fig. 1 Boundaries of the SZ SS ASS without taking into account the dependence of the TS on CA, for the condition of IHM (SS course $K_{SS} = 0°$)

(a)　　　　　　　　　　　　　　　　　　　　　　　　　　　　　**(b)**

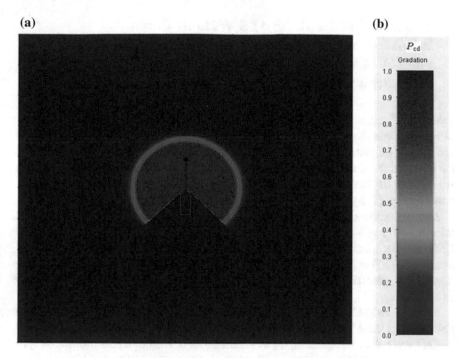

Fig. 2 Model SZ SS ASS without taking into account the dependence of TS on CA, for the condition of IHM, for arbitrary values $P_{cd} = 0$–1.0, taking into account the OCD of the sonar receiver (course SS $K_{SS} = 0°$)

the «cut out» sector of «non-listenable» stern CA, characteristic of SS ASS with keel-mounted array (Fig. 1).

In Fig. 1, SZ ASS is presented in polar coordinates (bearings are expressed in degrees, and distances are shown in kilometers), taking into account the real surveillance sector of the SS ASS in the horizontal plane CA $= \pm 135°$ and the course $K_{SS} = 0°$.

Figure 2a shows the SZ ASS model obtained for the same conditions and source data, but in the form of a «continuous» two-dimensional PCD function depending on the coordinates $P_{cd}(r, \alpha)$ taking into account the sonar receiver operating characteristics of the detection (OCD).

In this case, the visualization of arbitrary calculated values of PCD $P_{cd} = 0$–1.0 is carried out using a special scale P_{cd} (Fig. 2b), expressed as a color gamut (from blue-violet $P_{cd} = 0$ to red $P_{cd} = 1.0$), convenient for presentation three-dimensional graphs in the form of flat projections.

In Fig. 2a, the search area is represented as a square with a side of 80 km, the SS-carrier ASS is located in the center of the area, the course SS $K_{SS} = 0°$.

3   The Model of the SZ SS ASS Without Taking into Account the Dependence of Sonar Range on the TS and CA of Its Irradiation (for Conditions of a Layered Inhomogeneous Medium)

Taking into account, the influence of a LIM with boundaries (surface and bottom of the sea) leads to a more complex dependence of the type of SZ ASS. The shape and area of SZ will depend on the vertical distribution of the sound speed (VDSS), the depth of the sea in the search area, the depth of the antenna of the SS ASS, the submersion depth of the target, the state of the sea surface, the type of soil and the bottom relief [1, 2].

Figure 3 shows, as an example, the results of modeling of the typical modern SS ASS on a submarine (target) with a fixed sonar visibility (R_e = 10 m, TS = 13.98 dB) for specific hydrological and acoustic conditions (HAC): the Barents Sea, sea depth—200 m, season—winter, type of VDSS—positive refraction, the surface duct (Fig. 4), for different depths of immersion of the targets (submarine)—from 25 to 175 m.

Search area in Fig. 3 is presented in the form of a square with a side of 80 km, the SS-carrier of the ASS is located in the center of the area, the course of SS K_{SS} = 0°.

On each layer corresponding to a certain depth of the target, SZ represents symmetrical circular (ring) figures.

4   Dependence of Sonar Visibility (Target Strength) of Submarines on Course Angle of Its Irradiation

A typical view of the dependence of the sonar visibility of a submarine on the CA of irradiation α (in polar coordinates) is shown in Fig. 5.

Figure 5a shows the dependence of the radius of the equivalent sphere R_e (m) of the submarine; Fig. 5b shows the dependence of the target strength (TS) (dB) [1, 2].

It is seen that the graphs $R_e(\alpha)$ and TS(α) have axial symmetry with respect to the line CA = 0°–180°. In this case, the maximum values of the sonar visibility of submarines R_{emax} take place on the directions CA = ± 90° and the minimum values R_{emin}—in the stern CA = 180° and the nasal CA = 0° directions.

In the hydroacoustics literature [1], graphs TS(α) (dB) of the form (Fig. 5b) are figuratively called «butterfly», although in our opinion, in their appearance, these graphs more closely resemble a «flying mouse».

The blue dotted lines in the graphs (Fig. 5) denote «average» values of R_e = 10 (m) and TS = 13.9 (dB), which are most often used in traditional methods for approximate estimation of the active sonar range without taking into account their dependence on CA of irradiation.

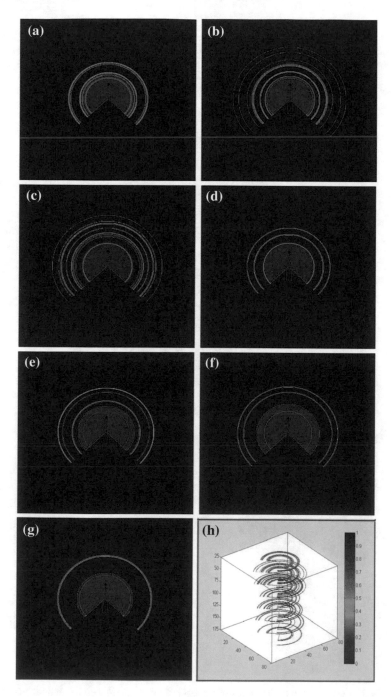

◀**Fig. 3** Model SZ SS ASS without taking into account the dependence of TS on CA, for the conditions of the LIM, for arbitrary values $P_{cd} = 0$–1.0 (course of SS $K_{SS} = 0°$) for different depths of immersion of the underwater targets H_T: **a** $H_T = 25$ m; **b** $H_T = 50$ m; **c** $H_T = 75$ m; **d** $H_T = 100$ m; **e** $H_T = 125$ m; **f** $H_T = 150$ m; **g** $H_T = 175$ m; **h** the representation of the volume SZ SS ASS (for possible values of H_T from 25 to 175 m)

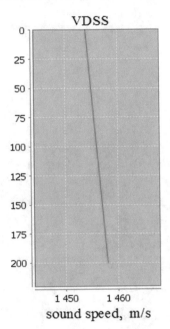

Fig. 4 Vertical distribution of sound speed (VDSS)

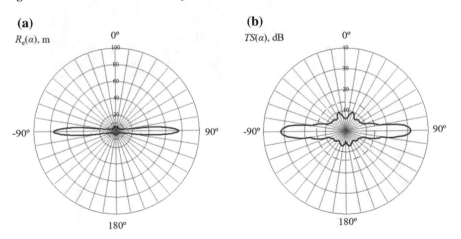

Fig. 5 Typical dependence of the radius of the equivalent sphere R_e (m) (**a**) and target strengths TS (dB) (**b**) for submarine on the course angle of its irradiation

$D_0(R_e)$, km

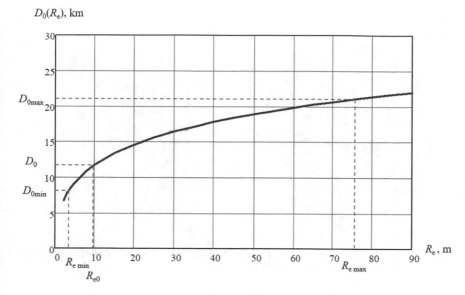

Fig. 6 Dependence of the sonar range in IHM $D_0(R_e)$ (km) on the radius of the equivalent sphere R_e (m) of the sonar target (for speed of SS-carrier ASS 12 knots, sea state three points)

5 Analysis of the Dependence of the Sonar Range in IHM on the Target Strength

The analysis of the dependence of the sonar range in IHM D_0 on the radius of the equivalent sphere R_e (m) for a typical modern SS ASS (Fig. 6) shows that this dependence is close to a power law of the form $D_0(R_e) \approx 5.5R_e^{0.31}$.

At the same time, for various possible values of R_e (m), ranging from ≈ 4 to ≈ 75 m, typical for sonar targets of the class of submarines (Fig. 5a), and, accordingly, target strengths (TS)—from 10.88 dB to 37.62 dB (Fig. 5b), with all possible changes in the CA of its irradiation, the sonar range in IHM of the typical SS ASS $D_0(R_e)$ varies from $D_{0\min} = 8.1$ km to $D_{0\max} = 21$ km (approximately 2.6 times from the minimum value to its maximum possible value).

6 Models of the Submarine Unmasking Zone and the Surveillance Zone of the ASS Taking into Account the Dependence of Sonar Range on the TS and CA of Its Irradiation (for Conditions of an Infinite Homogeneous Medium)

Taking into account the obtained dependences $R_e(\alpha)$ (Fig. 5a) and $D_0(R_e)$ (Fig. 6), we can find a new dependence of the sonar range on the CA of target irradiation $D_0(\alpha)$

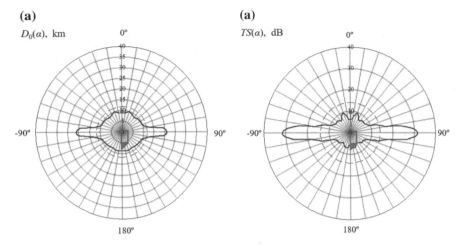

Fig. 7 Dependence of the sonar range in the IHM $D_0(\alpha)$ (km) on the CA of target irradiation (**a**) for a submarine with a target strength TS(α) (dB) (**b**)

$= D_0(R_e(\alpha))$ and build an appropriate diagram of the dependence of the sonar range on CA of target irradiation $D_0(\alpha)$ (Fig. 7). So, an idealized model of the submarine unmasking zone (UZ) in the sonar mode (on the returned underwater acoustic field) can be obtained for the conditions of an IHM—for a particular type of ASS installed on the corresponding type of SS-carrier moving at a given search speed.

Figure 7a shows the results of modeling (estimation) of the submarine UZ (according to its returned underwater acoustic field), taking into account the corresponding dependence of the TS(α) on the CA of target irradiation (Fig. 7b).

The red solid line in Fig. 7a denotes the boundaries of the obtained submarine UZ, taking into account the dependence of the TS(α) on the CA of the irradiation (Fig. 7b, red line).

Blue dotted line in Fig. 7a denotes the traditional approximate UZ $D_0(\alpha) = \text{const} = 11.7$ km for a constant target strength without taking into account its real dependence on CA irradiation; Fig. 7b is a simplified diagram of TS(α) = const = 13.98 dB, corresponding to $R_e(\alpha) = \text{const} = 10$ m.

The corresponding model of SZ SS ASS taking into account the dependence of the TS on CA, for the condition of IHM, (course of SS and target $K_{ss} = 0°$, $K_T = 0°$) is shown in Fig. 8.

In contrast to the submarine UZ, the SZ ASS, constructed with respect to SS, is a centrally symmetrically reversed figure, i.e., sonar range value corresponding to the direction of observation CA $= 0°$ (for the submarine following the course $K_T = 0°$), for SZ corresponds to the TS for CA $= 180°$ and vice versa.

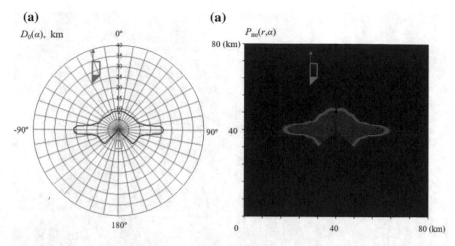

Fig. 8 Model of SZ SS ASS taking into account the dependence of TS on CA, for the condition of IHM (course of SS and target $K_{ss} = 0°$, $K_T = 0°$): **a** in the form of an SZ boundary with a fixed value of the PCD, equal to $P_{cd} = 0.9$; **b** for arbitrary values $P_{cd} = 0-1.0$, taking into account the OCD of the sonar receiver

7 The Model of the Surveillance Zone of the SS ASS Taking into Account the Dependence of the Sonar Range on the TS and CA of Its Irradiation (for Conditions of a Layered Inhomogeneous Medium)

Obviously, the SZ ASS model, taking into account the dependence of TS on CA irradiation, becomes unbalanced and dependent both on the target course (orientation of the TS diagram) and on the SS course (the position of the «non-listenable» CA stern sector). The change in SZ SS ASS (taking into account the dependence of the TS on CA irradiation) depending on the target course (on a plane, at a fixed immersion depth $H_T = 50$ m and a fixed SS course $K_{SS} = 0°$) is illustrated in Fig. 9a–f (target's courses $K_T = 0°$, $30°$, $60°$, $90°$, $135°$ and $180°$).

Blue dotted line in Fig. 9 indicates the direction of relative movement of the target relative to the SS-observer.

At the same time, the SZ ASS for the conditions of the LIM also depends on the depth of the target, as illustrated in Fig. 10a–h. Hydrological and acoustic conditions and initial data used in the modeling are completely similar to those given in paragraph 3 (VDSS—Fig. 4).

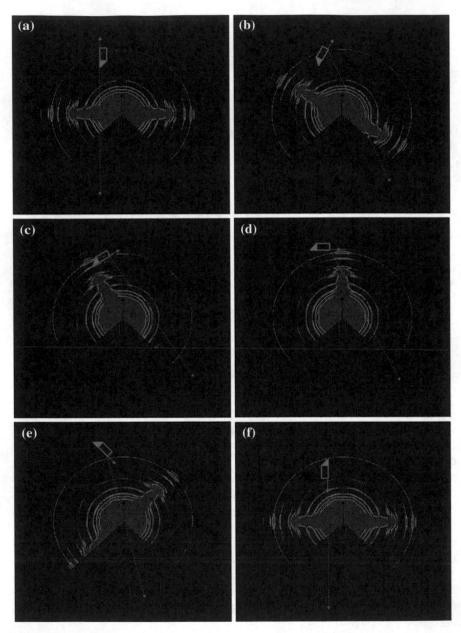

Fig. 9 Change in SZ SS ASS (taking into account the dependence of the TS on CA irradiation) depending on the target course K_T (on the plane, with a fixed target immersion depth $H_T = 50$ m and a fixed SS course $K_{SS} = 0°$): **a** $K_T = 0°$; **b** $K_T = 30°$; **c** $K_T = 60°$; **d** $K_T = 90°$; **e** $K_T = 135°$; **f** $K_T = 180°$

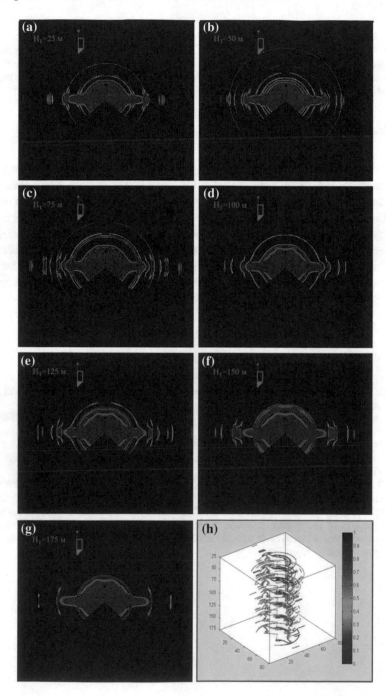

◄**Fig. 10** Model of SZ SS ASS taking into account the dependence of TS on CA, for the conditions of LIM, for arbitrary values $P_{cd} = 0–1.0$ taking into account the OCD of the sonar receiver (SS course $K_{SS} = 0°$) for different depths of the immersion of the underwater targets H_T: **a** $H_T = 25$ m; **b** $H_T = 50$ m; **c** $H_T = 75$ m; **d** $H_T = 100$ m; **e** $H_T = 125$ m; **f** $H_T = 150$ m; **g** $H_T = 175$ m; **h** the representation of the volume SZ SS ASS (for possible values of H_T from 25 m to 175 m)

8 Conclusion

The analysis of the obtained results of modeling of SZ ASS, taking into account the dependence of the TS on the irradiation CA, shows that the required accuracy of estimating the real SZ ASS, taking into account all environmental factors, as well as the difficult (complex) dependence of the SZ on the mutual CA of the ASS carrier and target can be received only with the use of modern GIS containing the necessary databases of the VDSS, as well as hydrological characteristics and bottom relief, for various trajectories of propagation of hydro location and echo signals [3–5]. Such an analysis of changes in SZ ASS is especially important in the process of direct mutual maneuvering of the ASS carrier and the sonar target.

References

1. Urick RJ (1975) Principles of underwater sound for engineers, 2nd edn. McGraw-Hill Book Company, New York, p 342
2. Burdic WS (1984) Underwater acoustic system analysis. Prentice-Hall, Englewood Cliffs, NJ, p 445
3. Popovich VV, Yermolaev VI, Leontyev YuB, Smirnova OV (2009) Modeling of hydroacoustic fields on the basis of an intellectual geographic information system (in Russian). Artif Intell Decision Making 4:37–44
4. Guchek VI, Yermolaev VI, Popovich VV (2012) Systems of monitoring on the basis of IGIS (in Russian). Defensive Order 2(21):58–61
5. Malyj VV (2017) Modeling of Surveillance Zones for Bi-static and Multi-static Active Sonars with the Use of Geographic Information Systems. In: Popovich VV, Schrenk M, Thill JC, Claramunt C, Wang T (eds) Proceedings of 8th international symposium IF&GIS'2017, Springer, Shanghai, pp 139–152

Analysis of Optical Images of the Sea Surface in the Interests of Environmental Monitoring

Andrey Grigoriev, Filipp Galiano, Maria Zarukina and Vasily Popovich

Abstract In recent years, the issues of environmental monitoring of the sea surface have become extremely relevant. The complexity of sea surface monitoring is determined by a number of key factors, major of which is the rapid variability of the marine situation in time caused by specific environmental conditions along with the impact of many factors, both natural and artificial. Such variability determines the order of conduction of experimental studies, and the methodology for processing of data obtained from geo-information systems (GIS). A large number of mathematical models and algorithms are used for data processing, and the record of dynamic changes of the environment in the investigated sea region according to GIS data is taken into account. This paper is devoted to solving of the pressing environmental monitoring issues.

Keywords Image processing · Environmental monitoring · Sea surface

1 Introduction

Monitoring of the environment for the purpose of ecological situation control is a challenging and complex task. Application of contact methods for environmental monitoring is often unsatisfactory according to the criteria of efficiency and the cost of obtaining results. The task of particular importance is the monitoring of the ocean surface, especially its coastal zone, in which three-quarters of the world's population reside. In this paper, two problems were considered: detection of anomalies on the sea surface and determination of the coastal zone's depth using remote sensing data. The

A. Grigoriev
Space Service Center "CosmoInform-Center" of Saint Petersburg State University of Aerospace Instrumentation, Saint Petersburg 190000, Russia
e-mail: grig-an@bk.ru

F. Galiano (✉) · M. Zarukina · V. Popovich
SPIIRAS-HTR&DO Ltd., Saint Petersburg 199178, Russia
e-mail: galiano@oogis.ru

importance of both tasks derived from the advantages of remote sensing over direct monitoring approaches: It has more speed and lower cost compared to traditional (direct) monitoring methods.

2 Analysis of Methods for Production and Processing Optical-Electronic Images of the Sea Surface

Water bodies' survey with application of optoelectronic means for environmental monitoring is characterized by the features listed further. Optical properties of water objects are non-stationary in the radiation spectrum. At some spectral intervals of the optical range, the water layer is transparent. At other spectral intervals, the water layer is non-transparent and only interacts superficially with the radiation. Water bodies are characterized by a large spatial coverage. This is typical for areal objects of survey. In some cases, it is necessary to monitor the coastal zone (the long-stretch object of survey).

The most productive optical-electronic means for obtaining images of the Earth's surface are based on aircrafts and spacecrafts. These tools can provide optical-electronic imaging with qualitatively different spectral and spatial parameters. The choice of specific parameters is determined by the properties of the researched or controlled objects. The specified parameters of the optical-electronic survey determine the information content of the recorded images in the framework of the applied problem.

Optical properties of water were studied in multiple theoretical works, for example [1]. Basically, water is characterized by high absorption capacity. There is an individual spectral range of water transparency in the blue-green part (450–550 nm) of the visible radiation (Fig. 1).

The statement about the transparency of the water layer in the blue-green spectral range is confirmed by full-scale experiments, which are performed using spectral optical-electronic imaging. Figure 2 shows examples of images that were obtained in narrow spectral bands. Figure 2 also shows image synthesized in natural colors.

The object of the research is a part of the water area, within which there is a contamination of the water layer with substances with mineral and organic compounds. Data analysis shows (Fig. 2) that the presence of contaminants is reliably recognized only in images, recorded in the spectral channels of the visible and near infrared ranges. With the transition to the short-wave infrared region (wavelength more than 1200–1400 nm), the absorption capacity increases and the visibility of anomalies in the water layer significantly decreases. Because of this, monitoring of contamination in the water layer is advisable to conduct in the visible and near infrared ranges. In the short-wave, medium-wave and long-wave bands, water layer is non-transparent. Optical-electronic survey in these ranges allows to explore only the properties of the water body's surface. These findings can come in handy for monitoring of surface contamination.

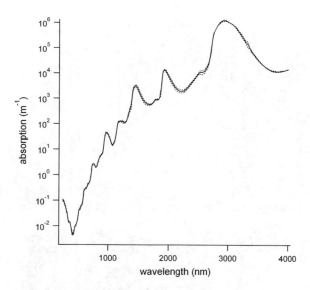

Fig. 1 Water absorption in spectral region 10–4000 nm [4]

The standard spectral parameters in optical-electronic survey include the number of spectral bands, the boundaries of the spectral bands and spectral resolution. Based on these parameters, methods for production of optical-electronic images are classified into panchromatic, multispectral and hyperspectral imaging [2]. All these methods of survey can be performed in visible and near infrared ranges of optical radiation. Methods of multispectral or hyperspectral survey that allow to identify the composition of water in detail can come in handy for solving problems of qualitative analysis of water areas and determining quantitative parameters.

Water bodies are diverse in their spatial properties. Water bodies in the optical-electronic survey in qualitative terms can be characterized by different spatial properties [3]. Multi-route survey methods are used to cover areal water bodies. Multi-route survey from an aircraft can be performed in the frame mode or in the scanning mode, from a spacecraft—mainly in the scanning mode.

In the context of solving the problems of environmental monitoring, it should be noted that the main sources of environmental danger for water bodies are grouped in the coastal zone. It is advisable to use the methods of curvilinear routes survey to examine the long stretch of coastal zone (Fig. 3). Such methods are relatively easy to implement with the assistance of aircrafts. This survey can be performed from the spacecraft after improvement of the onboard systems that provide control of the three-axis orientation. The long-stretch zone is split in separate parts that are observed in the azimuth mode shooting or in the shooting mode of the curvilinear route. Methods of photogrammetric processing are used for preprocessing of images, obtained in complex modes of survey of long-stretch and areal water bodies.

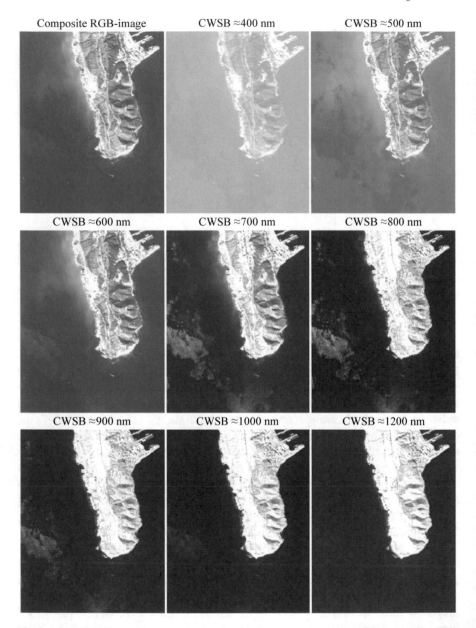

Fig. 2 Examples of images obtained in the spectral ranges with high spectral resolution. The captions of the images identify values of central wavelength of spectral band (CWSB)

(a)

(b)

Fig. 3 Examples of long-stretch object of survey (coastal zone). Result of photogrammetric processing of frame images, obtained after a survey of a curved route from an air-vehicle (**a**). Examples of azimuthal survey from space (**b**)

Methods of automated thematic processing of optical-electronic images of the sea surface can be applied for solving problems of environmental monitoring. These methods are widely promoted and implemented together with the use of multispectral and hyperspectral survey data. Methods of thematic processing of multispectral and hyperspectral images are classified into three groups:

– methods of target detection;
– methods of change detection;
– methods of anomaly detection.

The first group of methods is based on the use of identified stable features of the objects of interest (spectral, spatial and texture features) and allows to identify the spatial distribution.

Multitemporal images of the area of interest are required for the second group of methods [4]. Zones of change are determined on the basis of a joint multitemporal images analysis using a specified criteria. Zones correspond to the areas that have changed as a result of natural processes or anthropogenic activities.

The third group of thematic processing methods is used for the examination of area's objects with homogeneous optical properties. Zones, which differ significantly in their properties from the surrounding background, are identified based on the results of the spatial statistical analysis of the images. Such methods make it possible to detect and analyze shallow water areas in the context of the research of water bodies, as well areas with surface contamination or areas with the suspensions presence of mineral/organic substances in the water layer. Thus, the use of anomaly detection methods, based on optical-electronic survey, for environmental monitoring of water areas, is of particular practical interest.

3   A Generalized Image Processing Algorithm for Detecting Surface Anomalies and Estimation of Sea Depths

The remote sensing raster data processing software, developed by SPIIRAS-HTR&DO Ltd., is based on the approaches described in [5, 6] and allows to implement a wide range of raster image processing algorithms for solving tasks, presented in the previous section. The key features of the developed solution are:

1. Easy combination of analysis algorithms from a specified set, which makes it possible to create processes for customer's tasks.
2. Cross-platform; the ability to work on both the local machine and those distributed over the network.
3. The ability to easily add sets of algorithms in order to account for peculiarities of the problem at hand.

Using the developed solutions, the process of raster image analysis consists of the following three stages:

1. Preliminary processing of an image (averaging, filtration, changing of color space, etc.).
2. Detection of objects and calculation of their informative features (creation of feature space).
3. Classification of the objects using values of their features (properties).

To create the algorithm of raster analysis based on the developed solution for the problem of sea depth' estimation and detection of surface anomalies, it is necessary to take into account the following problem's properties:

1. Rapid change of the situation and, consequently, the need for initial and additional training of algorithms in the process of work.
2. Limited size of the training samples, which substantially limits the possibility of using some modern pattern recognition methods (neural networks, deep learning, etc.).
3. When using the software directly on board of an aircraft, it is important to take into account limited available computing resources.
4. Rapid change and dependence of the brightness and texture properties of the selected objects from the meteorological situation, time of day/year, etc.

In the process of research, two main tasks were solved:

1. Creation of an algorithm for detection of surface anomalies and
2. Detection of the sea depth in the coastal zone.

Both processes were characterized by the following common properties:

1. During preliminary processing stages, algorithms for changing color space and noise smoothing are used for increasing the robustness of the analysis [7].
2. An approach based on hierarchical segmentation using dynamic trees [8] was used for segmentation, because it improves the stability of the results in cases of changing meteorological conditions and provides high performance, which is especially important when processing data on board of an aircraft.

The first of the two tasks solved is the detection of anomalies on the water surface. By the anomaly, in this case, we mean areas that differ significantly in their texture, luminance or color characteristics from the normal sea surface. To isolate these kinds of anomalies, as a rule, algorithms for delimiting boundaries based on the calculation of a local gradient, or methods for image segmentation are applied. Experiments show that algorithms of the second type show greater stability in case of anomalies with fuzzy boundaries.

The second task is the estimation of sea depth, or, in other words, determination of lines of equal depth for ocean bathymetry. From a mathematical point of view, the problem can be formulated as a mapping of a set of pixels' brightness in different spectral ranges into a single number that is equal to the depth of the sea at the appropriate point on the ocean. In this phrasing, it is a particular case of the regression problem [7]. The main difficulty poses the dependence of the pixel's brightness from a set of environmental parameters, and not only from the actual depth of the sea. However, experiments have shown that it is possible to reduce the dependence on non-essential factors by preliminary converting the brightness of the pixels using preliminary processing. The stability of the results was ensured by using the following set of algorithms within the process:

1. As a preprocessing algorithm, the combination of color space change and smoothing in the spatial domain [7] was used.
2. As an algorithm that converts the pixels' brightness into the calculated sea depth, we use linear regression based on the training sample. The training sample was controlled by direct measurements of the depth, and the result was used to segment the selected zones of different depths on image.

As a case study, an example of division of ocean surface to the zones of equal depth is presented in Fig. 4. It shows the result of processing of small part of the coastal zone from Fig. 3a. Remote sensing data for this test was obtained with a visible-light spectrum camera mounted on a helicopter, from the zone of the Gulf of Finland. Lines with equal depth are drawn with yellow lines; calculated depth of each zone is marked automatically in zone's center using green numbers.

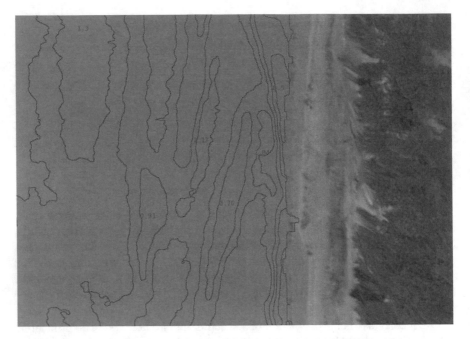

Fig. 4 Example of detecting sea depth using the data from the Gulf of Finland

4 Maritime Situation Consideration at the Instant of Surface Filming

In modern conditions, it is difficult to find an area of the sea or of water surface that bears no traces of antropogenic factors' influence, particularly traces of various kinds of water transport. Above-water objects on the sea surface manifest as both the monitored object and as the interference source, which can cause great difficulties for the analysis of sea surface and of interconnections between internal processes in water body and surface phenomena.

Figure 5 depicts an example of a typical above-water situation. It shows the offing of the Neva River, which is the most interesting area in the context of analysis. In order to avoid turning the sea surface analysis into scholasticism, monitoring of vessel traffic density in the water area under investigation should be performed in the most thorough manner at short time intervals. Only the complex analysis of vessel traffic and of specific vessels' tracks may allow to achieve objective estimations of sea surface analysis.

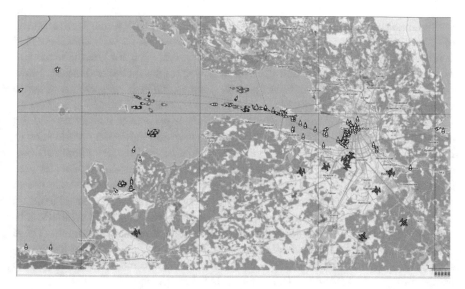

Fig. 5　The Gulf of Finland, offing of the Neva River, fairway

5　Interrelation of Water Surface Anomalies with Processes in Water

Among the causes for water surface anomalies may be the following:

1. slicks,
2. internal waves,
3. surfactants films (e.g., pollution of water with oil and products of its processing),
4. moving surface and underwater objects,
5. flowering of water,
6. water foaming, etc.

The anomalies of water surface are very different, have special features and different causes. Most anomalies are directly connected with the processes taking place in water.

Slicks are very common anomalies caused by wind waves. A wide variety of them is related to the structure of variable surface water currents, which, in turn, are associated with zones of waterfronts, internal and long surface waves, oceanic and near-surface vortices.

Surface water flows have both a direct effect on short gravitational-capillary waves and an indirect one—through the distribution of surfactants. Kinematic mechanism determines the transformation of surface waves by changing the wavelength and its intensity in the field of alternating flow. At the same moment, waves can both be strengthened and suppressed with the formation of slicks.

Oil pollution and other surfactants are clearly visible on the surface since the substances form a film. Unlike areas with clean water, additional attenuation of short gravitational-capillary waves takes place in areas covered with films. Redistribution of surfactants on the sea surface leads to the formation of areas with their increased concentration and reduced intensity of short wind waves. Film and kinematic mechanisms operate simultaneously and strengthen flow variables on the sea surface. Films can indicate not only the existence of pollution zones, but also the presence of zones of high biological productivity [9].

The concentration of various substances in water can affect its surface and the appearance of anomalies. The purity of water is characterized by the natural state of it without the influence of human activity. Any deviations in the water chemical composition outside of certain limits may negatively affect the condition of water area biota. Pollutants from sewage and effluents as well as any anthropogenic sources of emission affect the hydrochemical regime. Rivers bring to the ocean about 10 billion tons of waste annually. In addition to oil and petroleum products, toxic synthetic substances play a special part in the negative impact on the aquatic environment. These substances can form a layer of foam and the foam-producing ability in these substances occurs at a concentration of 1–2 mg/l. The concentration of these compounds in wastewater varies from 5 to 15 mg/l with MPC of 0.1 mg/l.

The change in the stratification structure of water can affect the anomalies of the water surface. The major types of changes in the stratification regime are [10]:

1. Disturbance of the stratification of waters of the Mictic and the Choric type,
2. Thermo-technogenic meromixis,
3. Chemo-technogenic meromixis.

The natural mix in nature also happens, but the mixing of adjacent layers occurs in this case. Mixing of disconnected layers of water is a distinctive feature of technogenic destratification, on the basis of which we can make the conclusion that anthropogenic interference is often the cause of water surface anomalies.

The anomalies of the ocean surface' temperature also belong to surface anomalies. Such anomalies most often occur in epicentral areas of strong underwater earthquakes, because correlation between the turbulent exchange coefficient and the acceleration amplitude of oscillating bottom is present near the water surface. The destruction of thermal stratification happens in regions of seismically active bottom. Temperature anomalies affect the near-water layer of the atmosphere. A noticeable change in temperature (one degree) of the ocean surface is possible from transfer of even a very small part of underwater earthquake energy (0.0001%) to the energy of turbulent motion. The almost complete destruction of thermal stratification will require 1–10% of earthquake energy, which is comparable to the energy of tsunami waves [11].

The Sargasso Sea surrounded on all sides by currents: the warm Gulf Stream, the Northern Trade winds and the cold Canary, is an example of anomaly. The borders of the sea are unstable and depend on season changes in the boundaries of currents. Moving clockwise, the currents cut off the cool waters of the North Atlantic from the Sargasso Sea so the water temperature in winter here does not drop below + 18 °C. The Sargasso Sea is a unique water area since it has no seashores. There is a large accumulation of brown algae (sargassa) due to the convergence zone of surface currents [12].

The application of statistical methods such as parametric, nonparametric and factor analysis is necessary to establish the relationship between anomalies recorded on the water surface and processes in water. Such method can be useful for detection of the correlation between processes and phenomena, for prediction and prevention of appearance of anomalies on the surface of water areas.

6 Conclusion

A solution for environmental monitoring of the sea surface was proposed in this paper. The proposed solution allows to monitor the sea surface using remote sensing methods, taking into account the physics of the processes occurring in the ocean, thereby potentially significantly improving the efficiency and accuracy of the analysis compared to direct (contact) methods or methods based on remote sensing without considering the physics of processes.

The anomalies of the water surface are different and can result from man-caused interference, oil pollution and products of its processing, high biological productivity, high seismic activity or peculiarity of natural conditions. The registration of meteorological data, physics and chemistry of processes is necessary for the detection, analysis and prediction of changes in water parameters.

Possible directions for further development of the proposed solution for analyzing raster images in the context of the addressed problem are usage of additional lidar data for creation of training samples and algorithms testing. Experiments show that green lidar can be used for determination of the depth of seabed in selected set of points, which can be used for improvement of depth calculation algorithms.

References

1. Ermakov S, Pelinovsky E, Talipova T (1982) Film mechanism of the internal waves action on the wind ripple. In the collection. Large-scale internal waves action on the sea surface (in Russian). Gorky. IPF Academy of Sciences of the USSR, pp 3–5
2. Grigor'ev A, Shilin B (2013) Analysis of seasonal variations of the spectral characteristics of landscape components, using the data of the Hyperion space video spectrometer. J Opt Technol 80:360–362
3. Grigoriev A, Dmitrikov G (2017) Spatial model and indicators of objects properties of remote sensing from space. In: Proceedings of CEUR workshop, vol 2033, pp 78–81
4. Röttgers R, Doerffer R, McKee D, Schönfeld W (2010) Pure water spectral absorption and real part of refractive index model ATBD (Draft by the University of Strathclyde Glasgov)
5. Galjano P, Popovich V (2007) Intelligent images analysis in GIS. In: Popovich V, Schrenk M, Korolenko K (eds) Proceedings of international workshop "information fusion and geographic information systems", 27–29 May 2007, Saint-Petersburg

6. Tsvetkov M, Galiano F (2017) Intelligent GIS-based remote sensing data analysis. In: Proceedings of CODATA 2017, Geoinformatics research papers, vol 5, bs1002. https://doi.org/10.2205/codata2017
7. Gonzalez RC, Woods RE (2008) Digital image processing, 3rd edn. Prentice Hall, Upper Saddle River
8. Galiano P, Kharinov M, Kuzenny V (2011) Remote sensing data analysis based on hierarchical image approximation with previously computed segments. In: Popovich V, Claramunt C, Devogele T, Schrenk M, Korolenko K (eds) Information fusion and geographic information systems. Lecture Notes in geoinformation and cartography. Springer, Berlin
9. Phillips O (1980) The ocean surface dynamics. Gidrometeoizdat (in Russian)
10. Suzdaleva A, Goryunova S (2014) Technogenesis and degradation of surface water objects (in Russian). Publishing House Energia, p 33
11. Zatsepin A, Khristoforova I (2014) The green resources of the Sargasso Sea—a renewable source of alternative energy. Successes of chemistry and chemical technology (in Russian). Book XXVIII, 4th edn.
12. Nosov M, Ivanov P (1995) The effect of a seismically active bottom on the stratification ocean structure (in Russian). Information Bulletin of the RFFI, 3rd edn.

Fusing Classification and Segmentation DCNNs for Road Feature Mining on Aerial Images

Lele Cao and Xin Pan

Abstract The availability of large amount of high-resolution aerial images, together with the recent advancement of deep convolutional neural networks (DCNNs) for extracting rich-and-hierarchical features from unstructured data, has propelled the automation progress of extracting roads from aerial images. Despite the superior performance of DCNNs, a common problem of choosing between the classification and segmentation DCNNs still remains. By comparing two state-of-the-art baseline classification/segmentation DCNNs in several industrial application scenarios, we illustrate that their relative performance may vary, leading to different choices. We also propose a strategy of fusing multiple pre-trained DCNNs and empirically discover that it guarantees superior results in all of the experimented scenarios, using far less development time. A few tools and pre-trained models (https://github.com/caolele/road-discovery) are open-sourced to facilitate research and engineering activities.

Keywords Road mining · Aerial images · Network fusion

1 Introduction

Have you encountered the situation where your navigation mobile application wants you to drive onto a nonexistent road? Before you burst into total madness and decide to uninstall that program, consider how ridiculously expensive and tedious it is to keep the road database accurate and always up-to-date without any scalable automated procedures? Therefore, a great deal of demands has arisen to automate those tasks. The fast development of high-resolution aerial imagery obtained from satellites and

L. Cao (✉)
Department of Computer Science and Technology, Tsinghua University, Beijing 100084, People's Republic of China
e-mail: caoll12@mails.tsinghua.edu.cn

X. Pan
Alibaba Group, Beijing 100102, People's Republic of China
e-mail: xin.px@alibaba-inc.com

© Springer Nature Switzerland AG 2020
V. Popovich et al. (eds.), *Information Fusion and Intelligent Geographic Information Systems*, Advances in Geographic Information Science,
https://doi.org/10.1007/978-3-030-31608-2_7

(a) Aerial image and label

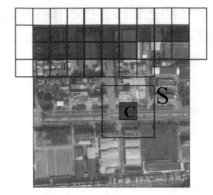
(b) Patch-based classification

Fig. 1 Illustration of **a** road annotations aligned with aerial images, **b** a classification DNN in action: each input tile S predicts a smaller patch C in its center

drones has made it possible to access huge volume of aerial imaging data; but the road maps are still constructed and updated mostly via costly and error-prone manual effort. So developing efficient approaches to perform fine-grained road extraction from aerial images has attracted more focus from fields such as computer vision and remote sensing.

The majority of the published researches dealing with automatic road extraction often rely on priori assumptions of prominent road characteristics and accordingly craft a multistep pipeline (e.g., [6, 15, 25]). Recently, the mainstream methodologies have shifted to learning road extraction from expert-annotated datasets (e.g., [7, 9, 11, 20]) since it is increasingly easier to obtain high-resolution aerial images together with aligned road labels (cf. the red polygon/polyline overlay in Fig. 1a). Therefore, many attempts (e.g., [16, 19, 22, 28, 29]) for predicting whether an individual pixel belongs to road area or not using certain pre-extracted contextual features have emerged. But those methods failed in generalizing to different scenarios. The state-of-the-art researches nowadays claim that applying an end-to-end learning architecture using deep convolutional neural networks (DCNNs) usually guarantees a superior performance than the aforementioned approaches. To the best of our knowledge, nearly all DCNN-based approaches fall into either **classification** or **segmentation** framework.

The road extraction accomplished via classification DCNN usually follow a so-called patch-based framework (e.g., [1, 8, 17, 21]) where the training images are cut into square patches **C** of size $w_c \times w_c$ as is illustrated in Fig. 1b. The objective is to predict the binary category for each pixel in **C** using a larger image tile **S** of size $w_s \times w_s$ where **C** locates exactly in the center. We also define $C_{(i,j)} = 1$ when pixel at coordinate (i, j) corresponds to road area and 0 otherwise. The classification DCNN is trained with a set of image tiles (i.e., S) to approximate a function f so that $f_{(i,j)} \in [0, 1]$ representing the probability that a pixel (i, j) in corresponding **C** is a road pixel;

Fig. 2 Segmentation DCNN has a form of encoder-decoder architecture where classification DCNN fulfills the encoder functionalities. FCs: fully connected layers

and pixel (i, j) is classified as road if $f_{(i,j)} > \sigma$ where $\sigma \in (0, 1)$. The common challenge of the classification DCNN mainly lies in (a) selecting a task-specific threshold σ and (b) joining the disconnected blotches in raw predictions.

Segmentation DCNN partitions an aerial image into two subsets of pixels which belong to either road or non-road class. To that end, segmentation may also be viewed as an approach of classifying the pixels of images, yet the fundamental distinction from classification DCNN is that its output must have the same size with input (i.e., $w_c = w_s$). That constraint requires all recent segmentation DCNNs follow a general encoder-decoder architecture, where the encoder part (left half of Fig. 2) is analogous to classification DCNN that extracts hierarchical feature representations, and the decoder part (right half of Fig. 2) approximates the full-scaled segmentation map from those learned features. Recent state-of-the-art segmentation approaches include fully convolutional networks (FCN) [24], U-Net models (e.g., [12]), LinkNet models (e.g., [32]), deconvolutional network (e.g., SegNet [2]), RefineNet [18], pyramid scene parsing network (PSPNet) [30], and DeepLab [3].

In this paper, we try to answer the practical question of "How to choose between a classification and a segmentation DCNN when building a real aerial imagery road extraction system for various industrial use cases? Or, do we really have to choose between them?" The contributions of this work are threefold:

- We experimentally showcase three industrial scenarios illustrating how the nature of tasks affects the choice of classification versus segmentation DCNNs.
- We empirically show that fusing the raw probability output from classification and segmentation DCNNs guarantees superior performance.
- Our fusion approach makes the life cycle of model development shorter since neither task-specific empirical study nor model pre-training is mandatory.

2 The Proposed Approach

We introduce a simple-and-effective approach that incorporates the strength of classification and segmentation DCNNs in different road feature mining tasks. The pre-trained classification and segmentation DCNNs is trained once and used everywhere, which makes the task-specific models much easier to be cooked for production applications.

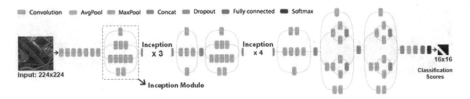

Fig. 3 A schematic illustration of GoogLeNet architecture used in our experiments

We adopt the GoogLeNet [27] (22 inception layers as illustrated in Fig. 3) as our classification DCNN. The inception layer acts as multiple convolution filters at different scales, that are applied to the same input, with some pooling operations; the results are then concatenated, allowing extracting multilevel feature representations in parallel. It also uses 1×1 convolutions that work like intra-channel feature selectors. With a carefully crafted design that allows for increasing the depth and width of the network while keeping the computational budget constant, the inference can be executed on individual devices including even those with limited computational resources and low-memory footprint.

For segmentation DCNN, we choose the DeepLab [3] model that applies the atrous convolution with upsampled filters for dense feature extraction. DeepLab also utilizes conditional random fields (CRFs) to obtain semantically accurate predictions and detailed segmentation maps along ROI boundaries. DeepLab obtained the state-of-the-art performance in several challenging benchmarks. We have noticed that it has been further extended to several variants such as [4, 5, 23], just to name a few. Due to the limitation of the production hardware, we employ the VGG-16 DCNN [26] instead of much deeper architecture like ResNet-101 [14]. We largely followed [31] to approximate the CRF layers (Fig. 4).

Recently in many other domains such as natural language processing (NLP), there are increasing researches on obtaining carefully pre-trained models and fusing the strength of them, some of which (e.g., [10]) obtained significant improvement over the state of the arts. Inspired by those approaches, we propose a fusion strategy (cf. Fig. 5) that stacks the probability map output from multiple pre-trained DCNNs.

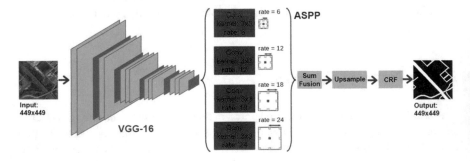

Fig. 4 VGG-16 DeepLab architecture (with CRF) used in our experiments

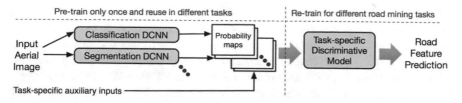

Fig. 5 Fusion strategy of classification and segmentation DCNNs

The stack further incorporates some task-specific auxiliary information (e.g., known road features), forming a multi-channel 3-D tensor. The 3-D tensor is further propagated into a task-specific model for predicting different road features. This paradigm requires only one careful pre-training of DCNNs; once pre-trained, they can be reused in different road feature mining tasks; the only part to be retrained/fine-tuned is the task-specific discriminative model.

3 Experimental Analysis

We pre-train DCNN models on three public datasets: DSTL satellite imagery feature detection [11], SpaceNet dataset for road detection and routing challenge [7], and DeepGlobe dataset for road extraction challenge [9]. We notice that while there are many high-resolution aerial imagery dataset available, per-pixel road annotations are hardly available because most of them merely specify the centerline of roads providing no information about road width. That is why we pick the three datasets specified below, which has either ready-to-use pixel-level road annotations or road labels that can be conveniently transformed to pixel-annotations covering the majority of road surface.

- DSTL [11]: We use the 3-band (RGB) 1 km(km) × 1 km satellite images (~0.3 m(m)/pixel) with "road" and "track" labels for training.
- SpaceNet [7]: The annotated training and testing images are 400 m × 400 m tiles (0.3 m/pixel); we use skeletonization and flood-fill to obtain a more accurate label mask for various kinds of roads (e.g., highways and alleyways).
- DeepGlobe [9]: We take the data for road extraction challenge; all samples are uniformly sampled from either rural or urban areas; to align with the above two datasets, every RGB image (19,584 × 19,584 pixels) are rescaled from 0.5 m/pixel to 0.3 m/pixel.

When training DCNNs, augmenting the training samples by rotating each input by a random angle is used to enhance the generalization ability of the trained models. Our implementation of GoogLeNet expects an RGB input tile of size 512 × 512 pixels that is then randomly rotated [0°, 360°) and cropped to 224 × 224 pixels; and DeepLab's raw input is 640 × 640 pixels that is cropped to 449 × 449 pixels after the same random rotation operation. So we cut each sample (from all three datasets regardless of the original image size) into small tiles (with stride $= 256$ pixels horizontally/vertically) of required input size from the two models. Considering that there are more than 90% negative pixels (non-road labels), we also discard the tiles without any labels to improve the balance of positive and negative pixels. Both the training and validation tasks are carried out on Nvidia Tesla K80 GPUs. The hyper-parameters (cf. Table 1) are tuned with a validation dataset containing about 10% randomly sampled image tiles.

To construct task-specific test datasets, we obtained high-precision road annotations together with aerial images (resized to 0.3 m/pixel) from a third-party mapping service provider; the selected areas embody drastically different geometric land-scapes: urban, rural, forestry, agricultural, highland, etc. We developed a Web-based road annotation tool (cf. Fig. 6) to adjust the labels for over 10,000 roads in such a way that (1) one polyline represents one road with constant width, (2) any road intersection splits a road into two polylines, (3) the width of any polyline must cover accurately the road surface.

Table 1 Hyper-parameters and validation performance of pre-trained DCNNs

Parameters and metrics	GoogLeNet (classification)	DeepLab (segmentation)
Initial learning rate	1×10^{-3}	5×10^{-4}
Learning rate decay[a] lr_policy: step	gamma $= 0.25$	gamma $= 0.65$
	stepsize $= 5 \times 10^4$	stepsize $= 1.5 \times 10^4$
	max_iter $= 5 \times 10^5$	max_iter $= 6 \times 10^5$
Momentum	0.9	0.96
Batch size	8 tiles	4 tiles
Validation performance	Precision $= \mathbf{75.10\%}$	Precision $= 73.24\%$
	Recall $= 79.35\%$	Recall $= \mathbf{82.60\%}$
	IoU $= 0.55$	IoU $= \mathbf{0.58}$

[a]The "step" decay policy of learning rate: http://caffe.berkeley/ision.org/tutorial/solver.html

Fig. 6 Screenshot of the road annotation tool for aerial images tailored to our needs

3.1   The Performance of Pre-trained DCNNs

The road probability map generated from the pre-trained GoogLeNet (e.g., Fig. 7b) is the spatial concatenation results of each 16×16 (Fig. 3) output patches; and that from the DeepLab (e.g., Fig. 7d) is assembled with 256×256-pixel patches that are essentially cropped from the center of 449×449-pixel DeepLab output (Fig. 4). To evaluate the pre-trained classification and segmentation DCNNs on the validation set, we adopt the metrics of pixel-wise precision/recall and Jaccard index (IoU: Intersection Over Union):

$$\text{Jaccard(IoU)} = \frac{1}{n} \sum_{\forall (i,j) \in C} \frac{y(i,j)f(i,j)}{y(i,j) + f(i,j) - y(i,j)f(i,j)}, \tag{1}$$

where $y(i,j)$ and $f(i,j)$ are the binary value (label) and predicted probability for the pixel located at (i,j) in the output map C, correspondingly.

In Table 1, DeepLab seems to have reached a better results in average, which might be attributed to the far more sparse classification labels (contain 16×16 pixels road map located in the center of the tiles) in the processed training dataset, making each classification sample pair less efficient in contributing to the convergence of the model. In the upcoming sections, we will further interpret that result by practically applying the trained model in different applications. We also noticed that our IoU scores are lower than the best results on the public leaderboard of each challenge, which we believe is caused by the reasons that (1) no post-processing in any form is used, (2) we only binarizing probabilities at a fixed threshold, (3) our dataset is a mix of three public datasets, (4) some label adjustment tricks are applied that might as well introduce new types of noise to the annotations, and (5) we did not exhaustively search for best hyper-parameters (incl. DCNN architectures) since this is not the primary objective of this research. We also empirically measured the relative temporal and spatial complexity for both DCNNs:

- DeepLab is almost 10 times faster than GoogLeNet to feedforward a same sized input;

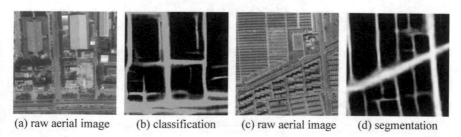

(a) raw aerial image   (b) classification   (c) raw aerial image   (d) segmentation

Fig. 7 Results of GoogLeNet (**a, b**) and DeepLab (**c, d**): the predicted road probability map without any post-processing or thresholding

- but the model size of DeepLab reaches 145 Mb (13 Mb after compressed [13]), while that of GoogLeNet is only 24 Mb (3.7 Mb after compression).

3.2 Performance of Predicting Road Width

The accurate information about road width is crucial for map service providers in, e.g., rendering road-network overlay and estimating the number of lanes. The objective of this task is to predict the width of roads provided the centerlines. For fair comparison, we apply the same approach on the direct output of GoogLeNet and DeepLab as illustrated in Fig. 8: We first obtain the tight square bounding box (i.e., a bbox of size $r \times r$) of the road centerline (located in the center of the bbox), which is then used to guide the cropping of the generated road probability map; the cropped map is resized to 384×384 pixels and stacked with the aligned road centerline in two different channels (three channels when fusing probability maps from both GoogLeNet and DeepLab); the stacks are then rotated ($90°$, $180°$, $270°$) and flipped (horizontally and vertically) resulting in six input samples that are respectively fed into a trained GoogLeNet regressor to obtain 6 scores (indicating the ratio between the road width and the diagonal length of input), the average of which is treated as the final prediction. The GoogLeNet regressor has a same architecture as Fig. 3 except that it ends with an FC layer with only one output; the regressor is trained, validated, and tested on 70%, 10%, and 20% of our dataset, respectively. We measure the performance (tenfold cross-validation) of both models using mean absolute error (MAE), root-mean-squared error (RMSE), and their standard errors (StdErr) in Table 2. The classification baseline performs better than segmentation baseline; and the fusion result approximately achieves another 30% improvement over classification DCNN.

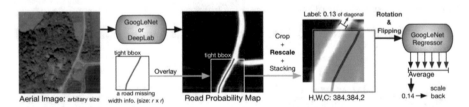

Fig. 8 Pipeline for predicting road width. Note that the input of the final regressor will contain three channels for the fusion case

Table 2 Performance on predicting road width

Fusion		Classification		Segmentation	
MAE	RMSE	MAE	RMSE	MAE	RMSE
0.022	**0.028**	0.032	0.039	0.045	0.050
(StdErr: 0.008)	(0.012)	(0.011)	(0.015)	(0.016)	(0.020)

3.3　Performance of Detecting Road Existence

Keeping the database of road network up-to-date is important to many Geographic Information Systems (GIS). Besides the traditional methods like ground surveying and vector map comparison, DCNN-based solutions using high-resolution aerial images have started becoming dominant recently because of the affordable acquisition cost of aerial imagery and the end-to-end feature learning capability requiring almost no prior domain knowledge. We hereby define this task as a binary classification problem to tell if a specific road (in the current database) still exists or not (1/0), by using only the aerial images. In real systems, it is highly inefficient to check each and every road on a regular basis; instead, it is a common practice to check only a small subset containing the low-heat roads (i.e., roads accessed with a much lower-than-usual frequency).

As illustrated in Fig. 9, the road probability output from GoogLeNet and DeepLab is stacked with the grayscale aerial image and the corresponding road polyline rendered with true width; they are also cropped loosely to encapsulate the entire road polyline, and resized to form a $384 \times 384 \times 3$ input for a binary GoogLeNet classifier (similar to Fig. 3, yet only has two output nodes). We augment the test sample by rotation and flipping (obtain six test samples including the original one), propagate them through the trained detector, and obtain the final result by performing majority voting over the six binary predictions. To obtain extra negative samples (non-existing roads), we downloaded the aerial images of the same areas from a much older period of time, pick the roads (from test dataset) that do not match the old aerial images, and add them into our negative dataset. The resulting dataset contains 14,200 roads. During the training of road-change detector, we apply random transformations to each sample, such as rotation, flipping, clipping, brightness/contrasts adjustment. We report the testing results in form of precision and recall in Table 3, which shows the segmentation DCNN appears to be more accurate in identifying the outdated roads, while the fusion approach provided another absolute uplift of about 2%.

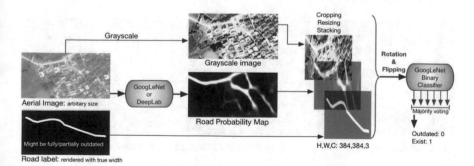

Fig. 9 Approach to detect road existence. Note that the input of the final binary classifier will contain four channels for the fusion case

Table 3 Performance on predicting road changes

Fusion		Classification		Segmentation	
Precision	Recall	Precision	Recall	Precision	Recall
0.972 ± 0.023	**0.933 ±** 0.024	0.932 ± 0.029	0.851 ± 0.032	0.956 ± 0.026	0.910 ± 0.030

3.4 The Performance of Generating Road Centerline

The attempt of directly converting the raw aerial images (e.g., Fig. 10a) to road vector format is challenging because of many difficulties such as varying image quality, overlapping features, and lack of metadata. However, by performing this task on the extracted raster maps representing road existence probability, the problem is greatly simplified and is fundamentally turned into a task of tracing along the centerline of strokes on road probability maps. As a baseline, we use an extension implementation in Inkscape[1] to generate road networks from output of GoogLeNet (e.g., Fig. 10b) and DeepLab (e.g., Fig. 10e).

To measure the difference between ground truth (e.g., Fig. 10d) and proposal road networks (e.g., Fig. 10c, f) from the perspectives of both logical topology as well as the physical topology of the roads, we adopt the metric of APLS (average path

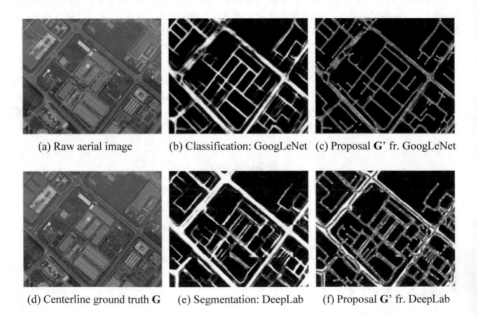

(a) Raw aerial image   (b) Classification: GoogLeNet   (c) Proposal **G'** fr. GoogLeNet

(d) Centerline ground truth **G**   (e) Segmentation: DeepLab   (f) Proposal **G'** fr. DeepLab

Fig. 10 Demonstration of generating road networks from raster probability maps

---

[1] Inkscape-centerline-trace: https://github.com/fablabnbg/inkscape-centerline-trace.

length similarity),[2] which sums the differences in optimal path lengths between all nodes in the ground truth network **G** and the proposal network **G'**. The APLS scales from 0 (poor) to 1 (perfect) and is calculated as

$$\text{APLS} = 1 - \frac{1}{N} \sum_{\forall(a,b)\in\{\text{none - paris in } \mathbf{G}\}} \min\left\{1, \frac{\left|l(a, b) - l(a', b')\right|}{l(a, b)}\right\}, \qquad (2)$$

where N is the total number of unique paths in G; Node a' is the same location in the proposal graph **G'** as the position of node a in graph **G**. $l(a, b)$ denotes a path distance in **G**, and $l(a', b')$ denotes the path length between the corresponding nodes in **G'**. Missing paths in **G'** are assigned 1.0. We observed that GoogLeNet and DeepLab achieved mean APLSs of 0.582 and 0.569, respectively; and the union of both DCNN probability maps (normalized) achieved a slightly better APLS score of 0.594. We also qualitatively observed that DeepLab tends to have more false activations of road areas, and adding some post-processing steps before performing centerline tracing may further improve the APLS score.

4 Conclusions

In this paper, we experimentally show that lacking the concrete contextual information of real application setups, it may be difficult to select between classification and segmentation DCNN approaches in various road feature mining tasks. Without any form of post-processing or exhaustive search of parameters, our segmentation baseline reaches slightly better IoU and recall on validation set. The classification baseline is much more accurate and stable in predicting road width, while segmentation baseline achieves a better precision and recall in the scenario of detecting outdated roads. For generating road centerlines, the two baselines perform almost equally, yet segmentation is more vulnerable to false road activations while classification is generally more computationally expensive. Nevertheless, we claim that it may not be mandatory to carry out such experimental analysis to choose between classification and segmentation DCNNs, since a simple fusion of pre-trained classification and segmentation DCNNs guaranteed superior performance in all the tested scenarios; it also coincides with many recent researches in other fields (e.g., [10]). Moreover, our approach makes the life cycle of model development shorter, since the DCNNs are trained only once and can be reused in many discriminative road mining tasks.

---

[2]APLS API based on Dijkstra's algorithm: https://github.com/CosmiQ/apls.

References

1. Aich S, van der Kamp W, Stavness I (2018) Semantic binary segmentation using convolutional networks without decoders. In: Proceedings of CVPR
2. Badrinarayanan V, Kendall A, Cipolla R (2017) Segnet: a deep convolutional encoder-decoder architecture for image segmentation. IEEE Trans Pattern Anal Mach Intell 39(12):2481–2495
3. Chen LC, Papandreou G, Kokkinos I, Murphy K, Yuille AL (2018) Deeplab: semantic image segmentation with deep convolutional nets, atrous convolution, and fully connected CRFs. IEEE Trans Pattern Anal Mach Intell 40(4):834–848
4. Chen LC, Yang Y, Wang J, Xu W, Yuille AL (2016) Attention to scale: scale-aware semantic image segmentation. In: Proceedings of CVPR, pp 3640–3649
5. Chen LC, Zhu Y, Papandreou G, Schroff F, Adam H (2018) Encoder-decoder with atrous separable convolution for semantic image segmentation. arXiv preprint arXiv:1802.02611
6. Christophe E, Inglada J (2007) Robust road extraction for high resolution satellite images. In: IEEE international conference on image processing, 2007, ICIP 2007, vol 5. IEEE, pp V–437
7. CosmiQWorks, DigitalGlobe, NVIDIA: SpaceNet on Amazon Web Services (AWS) Datasets: The SpaceNet Catalog. Accessed 3 June 2018. https://spacenetchallenge.github.io/datasets/datasetHomePage.html
8. Davydow A, Nikolenko S (2018) Land cover classification with superpixels and jaccard index post-optimization. In: The IEEE conference on computer vision and pattern recognition (CVPR) workshops, June 2018
9. Demir I, Koperski K, Lindenbaum D, Pang G, Huang J, Basu S, Hughes F, Tuia D, Raskar R (2018) Deepglobe 2018: a challenge to parse the earth through satellite images. In: The IEEE conference on computer vision and pattern recognition (CVPR) workshops, June 2018
10. Devlin J, Chang MW, Lee K, Toutanova K (2018) BERT: pre-training of deep bidirectional transformers for language understanding. arXiv:1810.04805
11. DSTL (2017) Kaggle satellite imagery feature detection. Accessed 3 June 2018. www.kaggle.com/c/dstl-satellite-imagery-feature-detection/data
12. Ghosh A, Ehrlich M, Shah S, Davis LS, Chellappa R (2018) Stacked U-Nets for ground material segmentation in remote sensing imagery. In: The IEEE conference on computer vision and pattern recognition (CVPR) workshops, June 2018
13. Han S, Mao H, Dally WJ (2016) Deep compression: compressing DNN with pruning, trained quantization and Huffman coding. In: Proceedings of ICLR
14. He K, Zhang X, Ren S, Sun J (2016) Deep residual learning for image recognition. In: Proceedings of CVPR, pp 770–778
15. Hu J, Razdan A, Femiani JC, Cui M, Wonka P (2007) Road network extraction and intersection detection from aerial images by tracking road footprints. IEEE Trans Geosci Remote Sens 45(12):4144–4157
16. Huang X, Zhang L (2009) Road centreline extraction from high-resolution imagery based on multiscale structural features and support vector machines. Int J Remote Sens 30(8):1977–1987
17. Kuo TS, Tseng KS, Yan JW, Liu YC, Frank Wang YC (2018) Deep aggregation net for land cover classification. In: The IEEE conference on computer vision and pattern recognition (CVPR) workshops, June 2018
18. Lin G, Milan A, Shen C, Reid I (2017) RefineNet: multi-path refinement networks for high-resolution semantic segmentation. In: IEEE conference on computer vision and pattern recognition (CVPR), pp 5168–5177
19. Mena JB, Malpica JA (2005) An automatic method for road extraction in rural and semi-urban areas starting from high resolution satellite imagery. Pattern Recogn Lett 26(9):1201–1220
20. Mnih V (2013) Machine learning for aerial image labeling. Ph.D. thesis, University of Toronto, Canada
21. Mnih V, Hinton GE (2010) Learning to detect roads in high-resolution aerial images. In: European conference on computer vision. Springer, pp 210–223
22. Mokhtarzade M, Zoej MV (2007) Road detection from high-resolution satellite images using artificial neural networks. Int J Appl Earth Obs Geoinf 9(1):32–40

23. Papandreou G, Chen LC, Murphy KP, Yuille AL (2015) Weakly-and semi-supervised learning of a deep convolutional network for semantic image segmentation. In: Proceedings of the IEEE international conference on computer vision, pp 1742–1750
24. Shelhamer E, Long J, Darrell T (2017) Fully convolutional networks for semantic segmentation. IEEE Trans Pattern Anal Mach Intell 39(4):640–651
25. Shen J, Lin X, Shi Y, Wong C (2008) Knowledge-based road extraction from high resolution remotely sensed imagery. In: Congress on image and signal processing, 2008, CISP'08, vol 4. IEEE, pp 608–612
26. Simonyan K, Zisserman A (2015) Very deep convolutional networks for large-scale image recognition. In: Proceedings of ICLR
27. Szegedy C, Liu W, Jia Y, Sermanet P, Reed S, Anguelov D, Erhan D, Vanhoucke V, Rabinovich A et al (2015) Going deeper with convolutions. In: Proceedings of CVPR. IEEE, pp 1–9
28. Wang J, Qin Q, Gao Z, Zhao J, Ye X (2016) A new approach to urban road extraction using high-resolution aerial image. ISPRS Int J Geo-Inf 5(7):114–126
29. Zhang Q, Couloigner I (2006) Benefit of the angular texture signature for the separation of parking lots and roads on high resolution multi-spectral imagery. Pattern Recogn Lett 27(9):937–946
30. Zhao H, Shi J, Qi X, Wang X, Jia J (2017) Pyramid scene parsing network. In: Proceedings of CVPR, pp 2881–2890
31. Zheng S, Jayasumana S, Romera-Paredes B, Vineet V, Su Z, Du D, Huang C, Torr PH (2015) Conditional random fields as recurrent neural networks. In: Proceedings of CVPR, pp 1529–1537
32. Zhou L, Zhang C, Wu M (2018) D-linknet: linknet with pretrained encoder and dilated convolution for high resolution satellite imagery road extraction. In: The IEEE conference on computer vision and pattern recognition (CVPR) workshops, June 2018

IGIS Algorithms and Computation Issues

Information Technologies: Modern Approach to Evolution of Methods of Obtaining Knowledge About Controlled Processes

Pavel Volgin

Abstract Development and practical application of objective research methods for investigation of pattern that can be found in the controlled processes play important role in maritime activities' management. In order to control, we must foresee. Ability to foresee comes only from learning about consistent patterns in the controlled processes. To realize the ability to foresee under modern conditions, one has to pay close attention to studies that allow to bring to light objectively conditioned, historically inevitable development trends in various fields of maritime activity, to determine the genesis of new theoretical and practical domains of management theory and to establish operating methods in new areas. Appearance of new methods is an objective aspect of advancement of methodology of obtaining knowledge about the controlled processes. Current stage of development of this methodology is inextricably linked to application of modern information technologies.

Keywords Maritime activity · Controlled processes · Modelling · Regularities · Ontology · Geoinformation technologies · Principles

1 Introduction

The process of implementation of modern information technologies, (especially geoinformation technologies, [1]) into all spheres of human activities is, undoubtedly, a process of monumental proportions. Naturally, this process was bound to influence such sphere as obtaining knowledge about consistent patterns in the controlled processes and a logical course of evolution of methods of obtaining knowledge about such patterns.

Scientific works [2, 3] by Professor Volgin N. S. illustrate the changes that occur in course of application and improvement of methods of obtaining knowledge about

P. Volgin (✉)
SPIIRAS Hi-Tech Research and Development Office Ltd., 14 Linia, 39, St. Petersburg 199178, Russia
e-mail: volgin@oogis.ru

© Springer Nature Switzerland AG 2020
V. Popovich et al. (eds.), *Information Fusion and Intelligent Geographic Information Systems*, Advances in Geographic Information Science, https://doi.org/10.1007/978-3-030-31608-2_8

113

consistent patterns in the controlled processes that have been utilized from the earliest times to the present day. These works provide an objective description of the role and value of methods of mathematical modelling and of electronics and computer technologies that can be regarded as precursors of modern information technologies. At the present time, the process of improvement of methods of obtaining knowledge about consistent patterns in the controlled processes has found its logical continuation in evolution of computer and information technologies. At the present time, further improvement of methods of obtaining knowledge about consistent patterns in the controlled processes proceeds on the basis of development of information technologies. The inevitability of improvement of such processes has been justified in [3].

2   Necessity of Learning About Consistent Patterns in the Controlled Processes

Under the controlled processes, we understand both processes of management over maritime activities and various structures and objects of maritime sphere of activity (Fig. 1) and the processes that are usual for scientific research and development tasks. In order to be specific, we will refer to management in the sphere of maritime activities when it does not interfere with the integrity of the discourse.

Patterns usual for maritime activities are depicted in form of principles and regulations used in maritime sphere of activity. Such patterns can not be deemed true or false. They are a part of objective reality. However, patterns can be perceived correctly or incorrectly, and, consequently, the corresponding principles and regulations will be true or false in maritime sphere of activity [2].

| Fishing activities | Drilling | Naval activities |
| Maritime transportation | Ocean exploration | Geophysical exploration |

Fig. 1 Various spheres of maritime activities

Patterns in maritime activities along with their reflections (principles and regulations) can address different areas of maritime activities (navigation safety, fishery, mining, armed struggle at sea, etc.) on various scales: local (areal), regional and global. They also can show different kinds of stability: they can hold true for a given historical epoch, can exist in the period of usage of some type of technological means or can be applicable only to realization of a process by means of a given set of tools and systems under the given conditions in presence (or in absence) of specific competitors.

But what methods allow us to obtain knowledge about consistent patterns in the controlled processes, to form principles and regulations of management activities? In order to answer this question, the process of forming and advancement of tools for learning about consistent patterns was examined in [2, 3] while drawing on the history of warfare.

Examination of processes in the historical context allows to identify their development trends. Following the stages of historical development of the society, especially the end of the eighteenth and the beginning of the nineteenth century, we can see that there were instances of complete absence of necessary experience in complex processes management, which was true not only for military activities but also for many other spheres of human activities. This has led to the appearance of a new research instrument: modelling. Modelling is used for obtaining knowledge about objects, events, processes and direct examination of which is for whatever reason impossible. It takes different forms when applied to various spheres of human activities. The common thing in all the cases is the reason for application of modelling: the impossibility of direct examination of the researched object (event) and, therefore, the necessity for creation of object's model and study of its properties through this model. The phases of research with application of modelling are also common, in general sense.

Opportunities that were opened by achievements in mathematics at the end of the nineteenth century along with the emerged need to compensate the limitations of the existing methods for identification of pattern induced emergence of another research tool: mathematical modelling (Fig. 2). The dotted rectangle in Fig. 2 depicts the endlessness of the process of improvement of methods of obtaining knowledge about consistent patterns in the controlled processes [3].

The distinctive feature of methods of mathematical modelling, the beginning of the extensive use of which in management practice dates back to the twentieth century, is the development of mathematical models of processes under study. One of the reasons why the methods of mathematical modelling have been widely implemented into management practice was a sharp contradiction between the need to increase the time of informed decision-making and realization plans' formulation on one hand and the need to decrease this time in order to increase the pace of realization of the controlled process on the other.

Implementation of mathematical modelling based on electronics and computer technologies and other technical means of automation of management at the end of the twentieth century has allowed to considerably alleviate this contradiction [2].

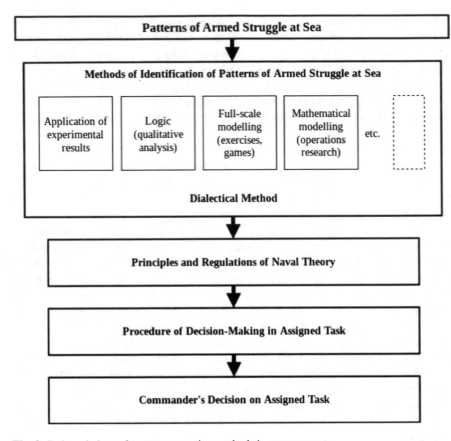

Fig. 2 Role and place of pattern perception methods in management

3   Contemporary Stage of Evolution of Methods of Obtaining Knowledge About Consistent Patterns in the Controlled Processes

Development of new methods of obtaining knowledge about consistent patterns in the controlled processes has never meant a renouncement of old methods. Rather, they underwent improvements through implementation of apparatus and techniques of new methods. At the same time, the possibility of application of previously existing methods was considered while forming new methods [3]. Application of mathematical modelling methods in management practice along with electronics and computer technologies has allowed:

- to come considerably closer to alleviation of the contradiction in management process mentioned above;
- to become the basis for appearance and wide implementation of modern information technologies in the management practice.

It would be no exaggeration to say that, in modern conditions, the application of information technologies has influenced practically every sphere of human activities. And the work of specialists related to the use of methods of obtaining knowledge about consistent patterns in the controlled processes, including such methods as qualitative analysis (logic), application of experimental results (Fig. 3, the dotted line depicts the limited scope of application of information technologies) is no exception. The wide implementation of information technologies is mostly due to application of modelling methods. Their application is carried out on all modelling stages: preparation, execution, result analysis and decision-making. However, information technologies are especially closely tied to methods of mathematical modelling and to software realization of any mathematical models.

Among methods of mathematical modelling, we can distinguish simulation methods the full capacity of which can be realized only with integration of information technologies into their creation and application [3]. Simulation modelling is one of the most widely used methods (quite possibly the most popular method) in operation research and in management theory [4]. Consequently, today, methods of obtaining knowledge about controlled processes are continuously evolving through their implementation into modern information technologies (Fig. 3) that are the most important material basis and component of management process.

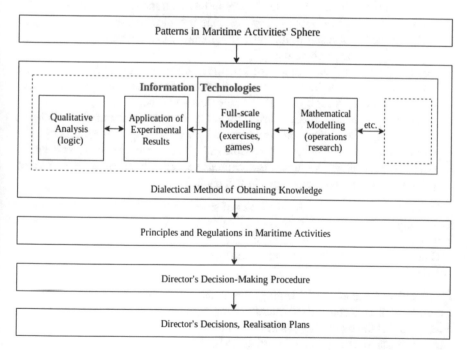

Fig. 3 Place of information technologies among methods of obtaining knowledge about patterns in management processes

Among modern information technologies that have a considerable influence on solving the above-formulated contradiction in management activities and have found wide practical application in various spheres of maritime activities, we can name the following:

- object-oriented modelling;
- ontologies;
- geoinformation technologies;
- information and telecommunication technologies;
- data and knowledge bases.

Information technologies provide wide spectrum of functional capabilities needed both for obtaining knowledge about consistent patterns in the controlled processes and for the management process as a whole, which provides necessary cyclicism and real-time realization.

Information technologies ensure implementation of various functions necessary for management and learning about consistent patterns in the controlled processes. Thus, object-oriented modelling (OOM) provides a number of advantages:

1. application of object-oriented approach significantly increases the degree of unification of development and suitability for reuse of created and tested models.
2. application of OOM allows to build systems based on stable intermediate descriptions which simplifies the editing process. This enables the system to evolve gradually and does not result in its full re-engineering even in case of considerable changes in initial specifications.
3. object-oriented models often turn out more compact. It means not only the decrease in code's volume but also reduction of expanses by using the previous developments which gives benefits in cost and time.
4. object model reduces risk of development of complex systems due to clearly defined stages of design and implementation of model creation process that stretches over the full development time rather than become a singular event.
5. object model allows to apply the expressive capabilities of modern object-oriented programming languages in full capacity.

Object form is best suited for the tasks of simulation modelling since it allows to definitively assigning the informational analogue to every object, event or process from the real world along with their relations.

Considering the whole spectrum of informational technologies, such functions include:

- possibility of expanding (changing) the modelling subject area and its contents;
- definition of concepts and terminology uniform for the subject area and used in (created for) the modelling system (ontology);
- full informational support of modelling and management processes (ontology, knowledge base and data base technology);
- temporal and spacial interpretation of real processes (geoinformation technologies);

- necessary level of visualization of simulated process and of modelling results (geoinformation technologies);
- access to remote informational and computational resources (information and telecommunication technologies).

Ontology system occupies an important place among modern information technologies used for obtaining knowledge about consistent patterns in the controlled processes. Ontology provides (Figs. 4 and 5):

1. opportunity to describe the subject area;
2. division of constant and variable information about objects. Use of the same object in different themes;
3. universal relation mechanism;
4. preservation of history about states of object's properties;
5. multiple inheritance;
6. filtering of information for different users.

Necessity of obtaining knowledge about consistent patterns in the controlled processes in maritime activities in real-time scale for a number of spacial processes, which usually compete among themselves, and which occur in sea areas (zones) with large spacial coverage, generates specific strict requirements for application software of mathematical modelling systems and, in particular, for application of geoinformation systems' technology. Contemporary level of requirements for the automation of the process of obtaining knowledge about consistent patterns in the controlled processes implies an automated support of decision-makers based on a set

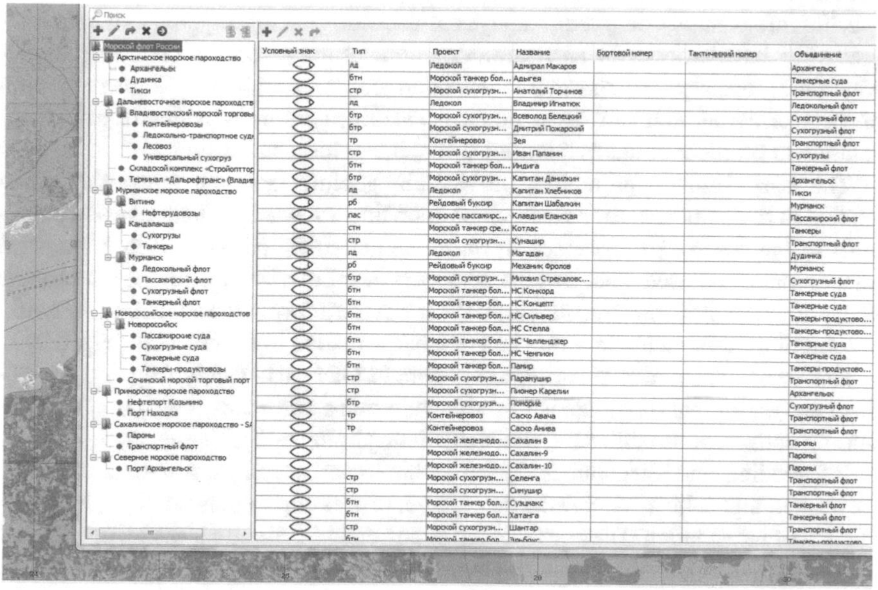

Fig. 4 Possibilities of subject area description. Universal relation mechanism

Fig. 5 Division of constant and variable information about objects. Use of the same object in different themes. Multiple inheritance

of mathematical models, i.e., formulation of recommendation on decision-making based on processing of object information obtained from various sources. At the same time, it is assumed that all applied software applications used for recommendation formulation are integrated among themselves; the recommendations are issued to user in a convenient (illustrative) form that requires minimum time for interpretation. That is why the geoinformation systems' technologies, into which the more specific information-computing, analytical decision-making support tools are integrated [5], have become the basis for automation of the process of obtaining knowledge about consistent patterns in the controlled processes. Moreover, implementation of geoinformation technologies in modern modelling systems provides a high-quality solution to problem of data's harmonization, integration and fusion.

The most critical opportunities provided by geoinformation technologies, when applied in mathematical models used for obtaining knowledge about consistent patterns in the controlled processes in maritime activities, are spacial and temporal integration of the used data, and the necessary level of visualization of the process under study. Geoinformation system (GIS) stores the information about the real world in form of a set of thematic layers and databases related to these layers (Fig. 6).

However, the sources of cartographic data in GIS vary widely (Fig. 7). Among them are traditional topographic and thematic maps and plans, field data, remote-sensing data, various table data, statistical data, photographs, video, etc.

Information and telecommunication technologies encompass various methods, means and algorithms for collection, storage, processing, representation and transmission of information. They provide the opportunity to utilize a large set of spatially distributed informational resources (Fig. 8) in the process of obtaining knowledge

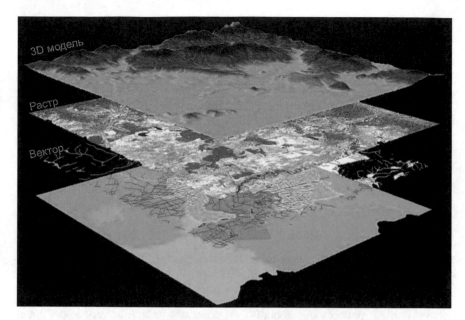

Fig. 6 An example of thematic layers of GIS

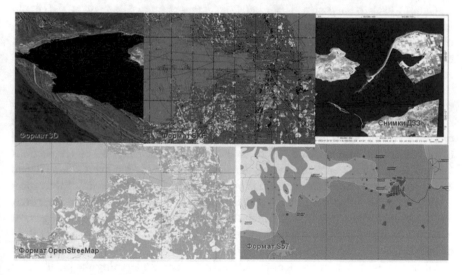

Fig. 7 Sources of cartographic data

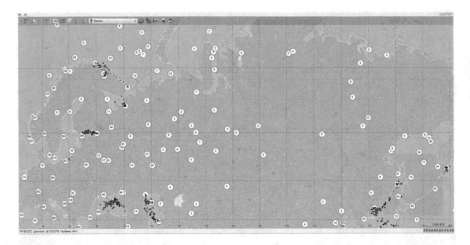

Fig. 8 Information about sea and river traffic obtained on-line from AIS and other maritime monitoring systems

about consistent patterns in the controlled processes, which is very important in analysis and management of various types of maritime activities.

Coming back to the problems of modelling, it must be noted that any model should meet contradictory requirements; on the one hand, the model should be similar to the object (event) understudy to the extent that the patterns under study are not distorted, on the other hand, it should differ from the object to the extent that it is possible to study its properties (as long as the object for any reasons can not be studied directly, it will be impossible to study its exact replica. You can not use the universe as a model of said universe! [3]).

At the same time, the first step of the research must entail preliminary examination of object's (event's) peculiarities. In the process of such examination, from a myriad of elements that compose the object or event and from a myriad of connections and interconnections between these elements, one must isolate such elements and such connections and interconnections which influence the properties of the object (event) under examination to the greatest extent. It is these elements, connections and interconnections that must be included in the model with necessary degree of adequacy to the real process which can only be achieved by involvement of all set of informational technologies.

Development and improvement of methods necessary for learning and justification of consistent patterns in the controlled processes is well-justified and historically inevitable. But familiarity with only these methods and even with patterns themselves does not guarantee an ability to develop rational solutions and plans. They are a product of leader's creativity. It could be argued that creative process also possesses its own patterns that were established by evolution involving the evolution of methods listed above.

Thus, it is evident that these patterns have different stability over time as well. Generally speaking, creativity of a person (leader, specialist), his style and heuristics[1] are in many ways a great secret. Recommendations, methodologies and methods that exist in this area base greatly on experience and gradually become outdated, undergo corrections and get redesigned. One of the indicators of need to improve the heuristics is emergence of new method of obtaining knowledge about consistent patterns in the controlled processes.

4 Conclusion

Summarizing the above, in can be affirmed that there exist two classes of patterns in the controlled processes:

- patterns typical for developed systems and for their functioning processes.
- patterns typical for human creativity involved in development of systems, in organization of their functioning and in their management.

These pattern groups are interconnected, interdependent, but they are not identical. There are methods for obtaining knowledge about said patterns which evolve and improve continuously. History demonstrates that improvement and development of new methods of pattern perception is a continuous process. Its primary cause is a continuous evolution of means and tools of production, progress of economy and society [3].

Human economic activity in the global ocean is very diverse (Fig. 1). For these activities, humanity has created and continues to create and upgrade high-performance tools and means of production, develops technologies and systems. However, their effectiveness is potential; it can be realized only under optimal management. Use of optimization results increasingly requires restructuring of the systems themselves: organizational, technical, mathematical models, etc. Under such high pace of management and execution of scientific research and design works in field of maritime activities, it is possible only when all set of informational technologies is put to use [6]. Therefore, it can be argued that this century will be a century of optimization and information technologies applied in process management not only in maritime activities but also in any other kind of human activity.

---

[1] Heuristics is a set of logic techniques and methodological rules of theoretical study and of discovery of some truth (solution). It is also organization of productive creative thinking that bears procedures aimed at solving creative challenges.

References

1. Popovich V, Pankin A, Voronin M, Sokolova L (2006) Analysis of situations based on intelligent GIS. In: Proceedings of MILCOM, Washington, USA
2. Volgin NS (1991) Systematic approach in the analysis of results of controlled processes, history and in forecasting. St. Petersburg, Russia. (in Russian)
3. Volgin NS (2002) Some thoughts on methodology of pattern perception in controlled processes and its evolution. Nauka, St. Petersburg, Russia (in Russian)
4. Law AM, Kelton WD (2000) Simulation modeling and analysis, 3rd edn. The McGraw-Hill Companies, Osborne
5. Popovich V, Claramunt C (2013) Distributed maritime observation systems for navigation. Proceedings of the ENC, Vienna, Austria
6. Volgin PN, Popovich VV (2018) Simulation modelling of maritime situation monitoring process. In: 14th All-Russia Conference "applied technologies in hydro-acoustics and hydro-physics", Russian Academy of Science, St. Petersurg, Russia (in Russian)

Model Analysis of Maritime Search Operation Using Geoinformation Technology

Alexander Prokaev

Abstract Traditional procedure for lost objects' search planning proposes development of probability distribution maps relies on the search object's possible behavior before his disappearance (because of accident, disaster, etc.). A group of experts develops one or several scenarios of object's possible behavior, and then these scenarios simulate using the Monte Carlo method. Simulation traditionally used for this purpose for more than half a century, and initial reason for that was low computational power of computers. Nowadays, it is possible to create quite complex scenarios based on analytical calculations using capabilities of modern geoinformation systems.

Keywords Search theory · Search planning · Maritime search operation · Scenarios of lost object's behavior · Geoinformation system (GIS)

1 Introduction

The very beginning of the search theory was primarily connected with needs of detecting German submarines in the Atlantic during the Second World War. In the postwar period, advances in theory and practice of maritime search operations were connected mostly with search of emergency objects sank because of accident or catastrophe. For instance, in 1966, the hydrogen bomb was lost in a plane crash near the coast of Spain. The bomb was detected due to the work of American scientists, Richardson [1] in particular, who made a map of the probability distribution for the location of a sunken object based on diverse and conflicting information from different sources. During the work, H. Richardson used scenario-based approach— the available data were combined into separate groups, so-called scenarios, and each of ones showed different picture of the lost object's position.

A. Prokaev (✉)
SPIIRAS—Hi-Tech Research and Development Office Ltd., St. Petersburg, Russia
e-mail: prokaev@oogis.ru

© Springer Nature Switzerland AG 2020
V. Popovich et al. (eds.), *Information Fusion and Intelligent Geographic Information Systems*, Advances in Geographic Information Science, https://doi.org/10.1007/978-3-030-31608-2_9

125

In June 1968, 400 miles southwest of the Azores, the US Navy nuclear powered submarine "Scorpion" sank because of the catastrophe. During the search of the "Scorpion," H. Richardson and L. Stone used a method of constructing probability maps based on the scenario approach applied by H. Richardson during the search of a hydrogen bomb in 1966 [2]. Debris of the submarine was detected at a distance of 240 m from the cell with the highest value of prior probability of detection.

Unfortunately, same accidents occur from time to time. One of the recent and deadliest catastrophes took place on March 8, 2014. Malaysia Airlines Flight MH370 was a scheduled international passenger flight operated by Malaysia Airlines that disappeared on 8 March 2014 while flying from Kuala Lumpur International Airport, Malaysia, to its destination Beijing Capital International Airport in China. It left radar range 200 nautical miles (370 km) northwest of Penang Island in northwestern Malaysia. With all 227 passengers and 12 crew aboard presumed dead, the disappearance of Flight 370 was the deadliest incident involving a Boeing 777 and the deadliest incident of Malaysia Airlines' history [3]. The search for the missing airplane, which became the most costly of aviation history, emphasized initially the South China and Andaman seas, before analysis of the aircraft's automated communications with an Inmarsat satellite identified a possible crash site somewhere in the southern Indian Ocean. Several pieces of marine debris confirmed to be from the aircraft were washed ashore in the western Indian Ocean during 2015 and 2016. After a three-year search across 120,000 km^2 (46,000 miles^2) of ocean failed to locate the aircraft, the Joint Agency Coordination Centre heading the operation suspended their activities in January 2017 [4]. A second search launched in January 2018 by the private contractor Ocean Infinity also ended without success after six months.

An opposite example of successful search operation was the search of Air France Flight 447, an Airbus 330–200 with 228 passengers and crew, disappeared over the South Atlantic during a night flight from Rio de Janeiro Brazil to Paris France. An international air and surface search effort recovered the first wreckage on June 6th five and a half days after the accident. More than 1000 pieces of the aircraft and 50 bodies were recovered and their positions logged [5]. The Phase II side looking sonar search, performed by the Pourquoi Pas from July 27, to August 17, 2009, proved unsuccessful. The Phase III search, which took place from April 2, to May 24, 2010, consisted of additional side looking sonar searches using REMUS AUVs operated by the Woods Hole Oceanographic Institute (WHOI) and using the ORION towed side looking sonar operated by the US Navy, was also unsuccessful.

In July of 2010, Metron Inc was tasked by the Bureau d'Enquêtes et d'Analyses pour la sécurité de l'aviation civile (BEA) to review the search and to produce an updated probability map for the location of the underwater wreckage [6]. To accomplish this, Metron reviewed and modified the previous prior distribution developed in 2009. The new prior distribution was based on studies by the BEA and the Russian Interstate Aviation Group (MAK) and a new reverse drift simulation with use of updated current estimates from the Drift Committee. Metron also analyzed the unsuccessful phase I and II searches performed during 2009, as well as the unsuccessful searches performed by REMUS and ORION in 2010 and including the photo and ROV searches. On April 3, 2011, almost two years after the loss of the aircraft, the underwater wreckage was located on the ocean bottom about 14,000 ft below the surface.

The history of successful search operations shows that the accuracy of knowledge of search object's location probability distribution determines success of the search at least in a half. The main method for search object's distribution modeling so far is mapping of the initial distribution and development of scenarios (accidents, disasters, etc.) of search object's "behavior" based on the opinions of several experts, followed by multiple statistical sampling of these scenarios (or final scenario, based on the consensus of expert opinion) using simulation based on the Monte Carlo method. Traditional application of simulation for creation of search objects' distributions one can explain by low computational power of computers and lack of computers.

According to our opinion, analytical model (when it is adequate and not too hard for creation) is much more profitable than simulation. There are many advantages that analytical models have over simulation ones. For instance, any simulation model is a so-called black box model—being activated one cannot adjust something before the moment when the model finishes its work. Analytical model is a "transparent" model—we can adjust everything what we want at any time during its work. This is especially important for real-time (or "close to" real time) models. The only traditional disadvantage of analytical model is its complexity; initially, we have to create mathematical algorithm, then to create computational one, and only then to create computer program. However, when we finally complete that, the reward will live up to our expectations.

Nowadays, computational power of modern computers and modern computer technology, such as math applications and geoinformation systems (GIS), allows to create fairly complex scenarios and to receive maps of prior (and then—posterior) search object's distribution based on analytical calculations in real time, especially in situations where uncertainty about location and activity of the search object is not deliberate.

2 Analytical Model of Search Object's Prior and Posterior Distributions

In the vast majority of real search situations, it is necessary to model the motion of the object along a curve, which is a result of complex influence of current and wind to the object's motion. Obviously, the curve can be approximated with a polygonal line. It is assumed that for each ith straight step, object moves over time t_i, courses in the sector $[\alpha_i, \beta_i]$ (in general, it can be equal to 2π), with the expectation of speed

u_i. Then the general expression for the calculation of the search object's density distribution after passing i steps can be written [7] as

$$f_i(x, y) = \int_{\alpha_i}^{\beta_i} f_{i-1}(x - u_i t_i \sin \varphi, y - u_i t_i \cos \varphi) d\varphi, \tag{1}$$

For conditions when route (or course) of the t search object is distributed according to the normal law with the value of its mathematical expectation on the ith straight step equal to α_i, and standard deviation equal to $\sigma \alpha_i$, expression (1) can be written as

$$f_i(x, y) = \frac{1}{\sqrt{2\pi}\sigma\alpha_i} \int_{-\pi}^{\pi} f_{i-1}(x - u_i t_i \sin \varphi, y - u_i t_i \cos \varphi) \cdot e^{-\frac{(\varphi-\alpha_i)^2}{2\sigma\alpha_i^2}} d\varphi \tag{2}$$

Expressions (1) and (2) can be transformed for different conditions of search object's motion; for instance, course of the object can have even distribution, mathematical expectation of speed u_i can have standard deviation equal to σu_i, initial distribution of the search object's position can be unconditioned, etc.

However, the computational experiment executed with the help of this model revealed a number of significant shortcomings of analytical methods for calculating prior probability and led to the conclusion that solution in the discrete form is much more profitable.

The essence of the analytical model for object's density distribution is as follows. In the analytical expression, calculation of object's motion performs by means of manipulation with the arguments of functions. In matrix expression, we perform similar manipulations, but with coefficients of matrix.

For example, the expression (1) in the matrix form is:

$$f_{k_{i,j}} = \frac{\Delta\varphi}{\beta_k - \alpha_k} \sum_{\varphi} f_{k-1}{}_{i-\text{round}\left(\frac{v_k t_k}{\Delta x} \sin \varphi\right), j-\text{round}\left(\frac{v_k t_k}{\Delta y} \cos \varphi\right)} \tag{3}$$

where

$\Delta\varphi$ is a step (i.e., interval) of calculating the course of the search object;

Δx, Δy are steps along the axes, respectively, x and y;

α_k, β_k are rates (respectively, minimum and maximum) of the search object's course on the kth step;

v_k, t_k are, respectively, speed and time of search object's rectilinear motion on the kth step;

round(x)—the operator defines the value of x to the nearest integer.

Thus, for the situation of discrete distribution, while moving an object from step to step:

- it is possible to modify the parameters of the search object's motion;
- no avalanche-like complication of the executing expression leads to the increasing of the calculation time.

However, this is not the only (though very important) advantage of the analytical model. Huge, not to say—a decisive, influence on the efficiency of the search has accuracy of environment parameters' knowledge, where the environment is a combination of different factors, influencing on the efficiency of the search. Data on the environment one usually receives in discrete form—it could be a digital raster map of search area or, for example, a weighted estimation distribution map of the emergency vessel location.

In this case, prior density distribution one can adjust considering parameters of the environment by means of element-wise multiplication of the initial matrix of prior density distribution to the matrix of the environment parameters distribution:

$$F_{\text{inv}_k} = F_k \cdot M_{\text{inv}} \tag{4}$$

where

F_{inv_k} is a matrix of prior density distribution taking into account the parameters of the environment;
M_{inv} is a matrix of environment parameters distribution.

Each element $m_{\text{inv}_{i,j}}$ of the matrix M_{inv} is the cumulative probability that the search object may be prior allocated in the specified cell. Given the need to comply with the condition of normalization, the matrix F_{inv_k} is to be normalized:

$$F_{k_{i,j}} = \frac{f_{\text{inv}_{k_{i,j}}}}{\sum_i \sum_j f_{\text{inv}_{k_{i,j}}}} \tag{5}$$

Thus, by means of the element-wise multiplication quite complex prior distribution of the search object, one can transform considering scenario of object's movement and parameters of the environment. Several scenarios of the search object's "behavior" (that mostly means movement) can take place. In this case, the final prior distribution F_k can be obtained by using the expression

$$F_k = \sum_n p_n F_{n_k} \tag{6}$$

where

p_n is the probability that the search object is moving in accordance with the nth scenario;

F_{n_k} is the matrix of the distribution after k steps of object's motion in accordance with the nth scenario.

Suppose that after kth step of the algorithm, this density is defined by matrix $F_k = (f_{k_{i,j}})$. Suppose further, that to the area X applied some search efforts, totaling Φ_0 so that the probability of object detection in each of the cells is $p_{i,j} > 0$, if the search took place there, and $p_{i,j} = 0$—if not. Thus, the probability of the event that the object is detected can be determined by matrix $P = (p_{i,j})$, and the probability of not detection—by matrix $Q = (q_{i,j})$, where $q_{i,j} = 1 - p_{i,j}$. Accordingly, the product $f_{k_{i,j}} \cdot q_{i,j}$ determines the probability of the event that the search object after completing the search is not detected, i.e., posterior density distribution.

As mentioned above, the distribution density matrix is to satisfy the condition of normalization. Given that the value of $q_{i,j}$ satisfies $0 \leq q_{i,j} \leq 1$, to find the posterior distribution density matrix of the search object $F_k^* = \left(f_{k_{i,j}}^* \right)$, it is necessary to apply Bayesian procedure

$$f_{k_{i,j}}^* = \frac{f_{k_{i,j}} \cdot q_{i,j}}{\sum_i \sum_j f_{k_{i,j}} \cdot q_{i,j}} \tag{7}$$

The search can be performed repeatedly in several steps; for example, a total of N times (and not necessarily at the end of the scenario, but at any of its k steps, with a probability of search object's detection: $P = (p_{n_{i,j}})$, where $n = 1, \ldots, N$). In this case, after each stage of the search matrix F_k^*, one can perform as follows

$$f_{k_{i,j}}^* = \frac{f_{k_{i,j}} \cdot q_{n_{i,j}}}{\sum_i \sum_j f_{k_{i,j}} \cdot q_{n_{i,j}}} \tag{8}$$

where $q_{n_{i,j}} = 1 - p_{n_{i,j}}$.

The value of the probability of object's detection because of the search efforts Φ_0 in a matrix form will look like

$$P_0 = \sum_i \sum_j f_{i,j} \cdot p_{i,j} = 1 - \sum_i \sum_j f_{i,j} \cdot q_{i,j} \tag{9}$$

where

$f_{i,j}$ is the object's density distribution at the start time of the search;

$q_{i,j}$ is an element of the matrix $Q = (q_{i,j})$ of the probability that the search object's is still undetected.

As we can see, all calculations use only element-wise operations with matrices. This type of calculations takes not such a big amount of computational power and operational memory as traditional analytical calculations do; for example, calculation of expressions (1) or (2), especially if number of steps i is large.

Below, there is an example of implementation of this method for modeling of the search object's distribution in a typical search situation using GIS capabilities.

3 Case Study

As an example, let us consider the use of the matrix method for modeling the distribution of a moving object, more precisely, emergency object on the surface of the water. The situation is that at some moment a message about an accident with the ship was received, the message included information of the ship's coordinates, and then communication with the ship ceased. According to experts' opinion, events after the accident could progress under one of the three scenarios:

- the ship remained afloat and was drifting under the influence of the wind and the current;
- the ship sank, but the crew was able to use collective survival equipment;
- the ship sank, but the crew was able to use only personal survival equipment.

 Standard error of the initial position of the ship, as well as the parameters of its motion in accordance with each scenario, based on the class of the ship and drift information in the search area, is to be defined by experts. This information we can get from different sources. It can be standard GIS information, average forecasts, and information received from the region of emergency in real time.

Drift, according to three scenarios, will differ one from another because of different contribution of wind and current to the object's movement. In the first scenario, wind and current move the ship together according to their direction and strength. The second scenario is mostly the same, but influence of wind and current will be different because the square of the above-water and below-water parts of the emergency object will be much less than in the first scenario. And, finally, the third scenario is rather similar to the second, but influence of the wind will be close to zero, influence of current will be much less than in the second one, and possibility to observe emergency object visually or by radar will be less than in the second and much less than in the first one.

In the area of the search operation, there are four radars; the type of a ship, radar antenna's height and the distance between the ship and the radar determine detection probability of emergency ship. In the search area, there is also the recommended way line, the intensity of traffic within a bandwidth of 5 miles from the specified

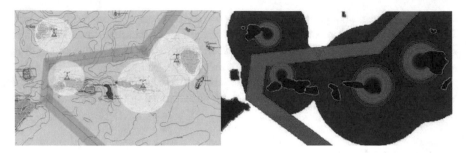

Fig. 1 Infrastructure of the search area (left) and the corresponding matrix of the environment parameters (right)

line is significantly higher than out of it; therefore, the probability of not detecting an emergency ship there will be lower.

The infrastructure of the search area is shown in Fig. 1 (left), and a matrix of the environment parameters corresponding to a given infrastructure and expert preferences is given in Fig. 1 (right).

Figure 2 (left) shows a prior probability distribution of the emergency ship location after 10 h movement according to three scenarios; using parameters of the environment (infrastructure of the search area), the ratio of probabilities of the scenarios was defined by experts as 5/2/1. In Fig. 2 (right), it is shown similar distribution at a ratio of probabilities of scenarios equal to 1/3/5.

If after some time of the search the emergency ship is still not found, the probability of its presence at different points of the search area has changed. Thus, density distribution of the search object after the search in the area is already called as posterior density distribution.

Fig. 2 Prior distribution of the emergency ship's location after 10 h movement according to three scenarios, and the ratio of the scenarios' probabilities is 5/2/1 (left) and 1/3/5 (right)

Map of posterior distribution is the basis for further optimization of the search by means of redistribution of search efforts. In the case of discrete distribution, the problem of posterior distribution modeling is solved as follows.

Figures 3 and 4 show the posterior distribution changing because of consistent survey of emergency ship's possible location areas as it moves according to three scenarios. The solid line indicates the position of already surveyed sections; dotted one is designated location of the next section of the search. Figure 3 (left) shows the prior distribution of the search object's in 6 h after the accident (the probability ratio of the scenarios defined by experts as 1/2/3, i.e., $p_1 = 0.167$, $p_2 = 0.333$, $p_3 = 0.5$) and position of the initial search section of the search area №1. Figure 3 (right) shows the posterior distribution after the search in the second section of the search area №1 and location of the third-search section in the search area №1.

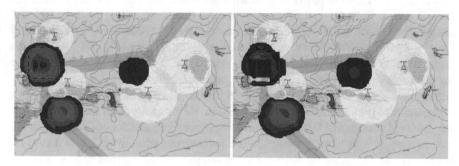

Fig. 3 Prior distribution of the emergency ship's location in 6 h after the accident and position of the initial section of the search aria №1 (left), the posterior distribution after the search in the second section of the search area №1 (right)

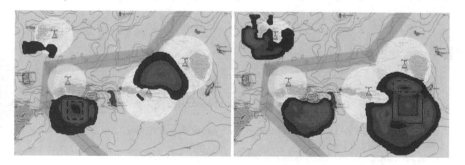

Fig. 4 Posterior distribution of the emergency ship's location after the search in the first section of the search area №2 in 8 h after the accident (left) and after the search in the second section of the search area №3 in 10 h after the accident (right)

Figure 4 (left) shows the posterior distribution after the search on the first section of the search area №2 in 8 h after the accident, and the probability ratio of the scenarios is 1/5/5, i.e., $p_1 = 0.091$, $p_2 = 0.455$, $p_3 = 0.455$. In turn, Fig. 4 (right) shows the posterior distribution after the search on the second section of the search area №3 (that corresponds to search object's movement according to scenario №3) in 10 h after the accident, and the probability ratio of the scenarios is 1/1/5, i.e., $p_1 = 0.143$, $p_2 = 0.143$, $p_3 = 0.714$.

4 Conclusions

The history of successful search operations shows that the accuracy of knowledge of search object's location probability distribution determines success of the search at least in a half. Currently, all Computer-Assisted Search Planning Systems (CASP), including the most advanced, for instance—adopted in 2007 the US Coast Guard System SAROPS [8], create mapping of the search object's probability distribution using simulation on the Monte Carlo method base. Simulation was traditionally used for this purpose because of low computational power and lack of computers. Today, practice of computer modeling of search actions convincingly shows that computational power of modern computers and modern computer technology, such as math applications and GIS, allows creating probability distributions of motionless and moving search objects using quite complex scenarios in real time based on analytical models. As a result, we have a number of advantages over simulation models, especially in real-time models.

In particular, an example of case study shown above, designed with the use of the considered analytical model based on GIS "Ontomap" developed by SPIIRAS—Hi-Tech Research and Development Office Ltd, St. Petersburg, Russia.

References

1. Richardson HR (1967) Operations analysis. Chapter 5 of part 2 in Aircraft salvage operation, Mediterranean In: Report to Chief of Naval Operations from the Supervisor of Salvage and Deep Submergence Systems Project, US Navy
2. Richardson HR, Stone LD (1971) Operations analysis during the underwater search for Scorpion. Naval Logist Res Q 18:141–157
3. https://en.wikipedia.org/wiki/Malaysia_Airlines_Flight_370
4. The Operational Search for MH370 (2017) Final report, Australian Government, Australian Transport Safety Bureau
5. Stone LD, Keller CM, Kratzke TM, Strumpfer JP (2011) Search analysis for the location of the AF447. Technical report, BEA
6. Stone LD, Keller CM, Kratzke TM, Strumpfer JP (2014) Search for the wreckage of Air France Flight AF 447. Stat Sci 29(1):69–80

7. Popovich VV, Hovanov NV, Ermolaev VI, Prokaev AN (2016) Theory of detection and search of mobile objects. Nauka, St. Petersbur, 424p
8. Frost JR, Kratzke TM, Stone LD (2011) Search and rescue optimal planning system. In: SaferSeas conference, 10 May 2011, Brest, France, pp 245–276

IGIS for Urban and Land-Based Research

Context-Driven Tourist Trip Planning Support System: An Approach and OpenStreetMap-Based Attraction Database Formation

Alexander Smirnov, Alexey Kashevnik, Sergey Mikhailov, Nikolay Shilov, Daria Orlova, Oleg Gusikhin and Harry Martinez

Abstract The number of tourists has significantly increased recently. In 2016, the total number of tourists in the world became more than one billion. People travel around the world and they are interested in new information systems that could save their time and money and provide additional context-related information about the location. Evolution of information and communication technologies and geographical information systems enables creating new information systems for tourists that provide them with a higher level of user experience. Today every tourist has a smartphone that can acquire information from various sensors and provide a comfortable interface to such information systems. The paper proposes an approach to the development of a tourist trip planning support system that is aimed at trip generation based on tourist's preferences and context information in the considered region. Since Internet access might not be available in some places and downloading large volumes of information abroad can be expensive, it is proposed to prepare the attraction database offline, download it to the user smartphone and utilize it during the trip.

A. Smirnov · A. Kashevnik (✉) · S. Mikhailov · N. Shilov
SPIIRAS, 39, 14 Line, St. Petersburg, Russia
e-mail: alexey@iias.spb.su

A. Smirnov
e-mail: smir@iias.spb.su

S. Mikhailov
e-mail: sergei.mikhailov@iias.spb.su

N. Shilov
e-mail: nick@iias.spb.su

S. Mikhailov · D. Orlova
ITMO University, 49 Kronverksky Pr., St. Petersburg, Russia
e-mail: orlovadasha@inbox.ru

O. Gusikhin · H. Martinez
Ford Motor Company, 2101, Village Rd., Dearborn, MI, USA
e-mail: ogusikhi@ford.com

H. Martinez
e-mail: hmartin3@ford.com

© Springer Nature Switzerland AG 2020
V. Popovich et al. (eds.), *Information Fusion and Intelligent Geographic Information Systems*, Advances in Geographic Information Science,
https://doi.org/10.1007/978-3-030-31608-2_10

For the database formation, it is proposed to use OpenStreetMap service to collect information about attractions and Wikipedia service for extraction of the media content about these. The prototype of the tourist trip planning support system has been implemented for Android-based smartphone and tested by a group of tourists in St. Petersburg. Furthermore, the system is capable to dynamically connect with vehicle infotainment systems to enhance the quality of interaction with the tourist.

Keywords Tourism · Context · Information and communication technologies · GIS

1 Introduction

Nowadays, the tourism industry has been increasing from year to year. In the 2016, the total numbers of tourists in the world have become more than one billion [1]. Development of the information systems that enhance the tourists' experience is important and actual at the moment task.

The paper is aimed at description of the developed context-driven tourist trip planning system. The system includes services for fetching information about attractions from OpenStreetMaps (OSM) GIS and Wikipedia and provides possibilities to enhance this information by tourism experts and to provide the information to the tourists. The system supports integration with vehicle infotainment systems that enhance the quality of interaction with the tourist. However, the system is not related to the augmented reality topic.

The following groups of applications have been identified based on travel phases:

- pre-travel phase, that provides range of services to facilitate travel-related information search, such as attraction descriptions, hotel, and airplane booking;
- travel phase, that provides the tourist with real-time information about the destination, e.g., information about events, places of interest, advices, and practical recommendations;
- post-travel phase, that is aimed at getting feedback from the tourist (there exists a variety of solutions to collect attraction estimations by tourists) and share his/her travel experience with others.

The first application group provides the tourist with a possibility to plan his/her trip, get information about attractions for a given destination, and book hotels and flights. The applications from the second group provide the tourist with personalized context-based information about attractions in the destination. The aim of the applications from the third group is to collect posts, photos, videos, and/or estimations of attractions attended by the tourist. This information can help other tourists to decide if they would like or not to attend these attractions. There are applications that cover functionality related to two or three groups. For example, TripAdvisor allows to plan the tourist trip by browsing information before the trip and to collect feedback after the trip.

The rest of the paper is as follows. Section 2 describes the general approach to the tourist trip planning system development. Section 3 describes services that provide functionality for the extraction of attraction data, the formation and manipulation of constructed information. Section 4 contains conclusion and the future work.

2 Related Work

The carried out analysis of the tourist support systems showed that they can be divided into two main groups for information extraction: (1) systems that implement information search in the Internet and (2) systems that have their own databases. Systems from the first group require an Internet connection while applications from the second group can provide the tourist information offline. Both types of systems utilize the recommendation algorithms to provide the tourist with the most interesting for him/her attractions. For example, collaborative filtering algorithms are aimed at estimation of the attraction (if it is interesting to the tourist and based on the previous tourist ratings and ratings of other tourists in the system including the ratings of the considered attraction). The most relevant tourist trip planning solutions are presented in Table 1.

Table 1 List of tourist trip planning solutions

Name and link	Description	Platform
XplorerVU [2]	The XplorerVU application analyzes the context situation by using tourist's social network profile and previous check-ins. Recommendations are made by using novel user clustering based on Quantum-behaved particle swarm optimization and d on the user personal activities for the urban POIs	Research
SpaceBook project [3]	SpaceBook utilizes hands-free and eyes-free virtual tour guide by using automatic speech recognition. This developed system can notify the user about attractions that are in his/her view while guiding the tourist to various destinations	Research

(continued)

Table 1 (continued)

Name and link	Description	Platform
UTravel smart mobile application [4]	The application is aimed to recommend point of interests (POIs) to the end user. A profiling and recommendation approach, based on context-awareness, is presented in the paper. The user profile in this application is built by taking into account not only individual behavior but also the behavior of other similar groups of users	Research
FM-based recommender [5]	The paper is aimed at recommendation generation for rural tourism. Authors propose effective methods of seasonal feature extraction and geographical distribution feature extraction	Research
The paper [6]	The paper describes a set of models and algorithms used in a tourism recommendation system based on user profiles and point of interest (POI) profiles. Also, this work presents an interesting approach to POI classification considering their accessibility levels, mapped with similar physical and psychological issues	Research
Go!Tour [7]	This is an Android-based mobile application for providing tourism and geographical services in Istanbul city. The application has internal attraction database and provides possibilities of searching places of interests around using the variable neighborhood algorithm	Android OS

(continued)

Table 1 (continued)

Name and link	Description	Platform
World Around Me [8]	This is a Windows Phone 7 application that shows the user photos around the user location. Photos are automatically downloaded from Flickr and Panoramio and presented to the user	Windows Phone 7
Triposo http://www.triposo.com/	The travel guide Triposo is a free mobile guide service available for Apple and Android devices. A user can download the application and appropriate database to the mobile device beforehand and use it during the trip without the Internet connection. The application supports logging of traveling. Each guide contains information on sightseeing, nightlife, restaurants, and more	iOS Android OS Nokia Ovi Store
TripAdvisor http://www.tripadvisor.ru	TripAdvisor provides for traveler reviews, photos, and maps. Tourists can plan their trips taking into account over 100 million reviews and opinions by travelers. TripAdvisor makes it easy to find the lowest airfare, best hotels, great restaurants, and fun things to do	iOS Android OS Nokia Ovi Store Windows Phone 7 Web application
ARTIZT [9]	ARTIZT is an innovative museum guide system, where a Zigbee protocol is used to determine user's position information. Visitors use tablets to receive personalized information and interact with the rest of the elements in the environment. The system achieves a location precision of less than one meter. The context is used to provide needed at the moment personalized information to the user	Prototype

3   Approach to the Tourist Trip Planning System Development

The proposed approach to the tourist trip planning system development is aimed at tour routes generation and tourist support (guiding, narration, and presentation of additional information) based on the personal schedule analysis, current situation monitoring and prediction, and integration of the on-board infotainment system with personal smartphone/tablet and external services via Ford's SYNC Applink API (see the overall scheme in Fig. 1). It is based on the earlier developed by the authors' personalized tourist attraction information service (TAIS) [10] and other authors' research done in this area [11, 12]. The tour is generated based on the preferences stored in the tourist's profile, situation in the area and its possible development (e.g., regular traffic jams during rush hours can be easily predicted), and available information about attractions.

Tourist(s) can use personal smartphones/tablets during the tour for narration, imagery, and video synchronized with the vehicle's location, speed, and orientation. Guiding information is extracted from accessible in the smart city services and predefined libraries.

The user is also able to communicate in some extent through the SYNC system with the driver if he/she does not speak the user's language (e.g., ask for a stop near an attraction) and control some car elements as opening/closing windows or adjusting climate control. The system takes into account the context information including tourist's preferences described in his/her profile (preset and revealed via collaborative filtering techniques) and the current situation at the location (season, weather, traffic jams, etc.) to anticipate what the passenger would want and need. All this information constitutes the context of the current situation that affects the tour flow and might cause changes in it. Here, the context includes both the environmental information and tourist information. The tourist information is described in the tourist

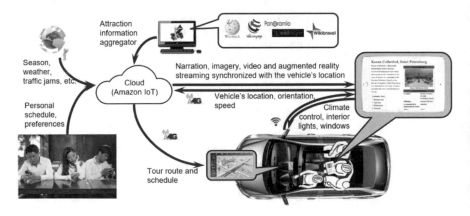

Fig. 1 Overall scheme of the on-board dynamic tour support system

profile and is considered separately further in the paper, and the context assumes only the environmental information.

For accumulating information about the attraction and providing it to the tourists, the trip planning system is based on a set of services (see reference model in Fig. 2). The data processing service is designed to gather data about attractions in the specific region. This information includes attraction name, coordinates, type ("museum," "monument," "theatre," etc.), general description, images, etc. This service uses the OSM service as the main source of geographical data and treats Wikipedia as the main information source for attraction fetching. After the data processing, all created information is packed into the attraction database.

The attraction matching service takes basic attraction information such as title and coordinates and tries to find the best suitable Wikipedia page that describes this attraction. The proposed page can describe the attraction itself or may be associated with a person or event. The Wikipedia API is used for page searching. The found results are filtered, and the best results are sent back to the data processing service for information extraction.

The attraction managing service is created in order to view and interact with the created regional databases. The experts can use the Web-based interface for creating new attractions and editing the existing ones in the selected regions. It is possible to add new media files to the attraction such as audio or video files.

The attraction delivery service stores the attraction database and provides the tourists with relevant attraction information at the right time. Information in the database is stored in the SQLite format and is suitable for offline usage on the user smartphone. The service acts as a Web service and sends data by POST request.

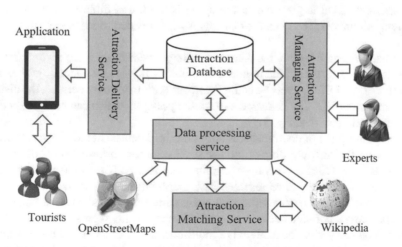

Fig. 2 Reference model of the tourist trip planning system

4 Service Description

4.1 The Data Processing Service

The data processing service fetches the attraction geo-based data from the OSM and matches the collected information with crowdsourced information storages in an attempt to create a complete attraction description. At the moment, Wikipedia project is treated as the main information source to obtain the attraction media details, such as text description, and images, but it is possible to use other information sources. All collected information is stored in object-relational mapping objects, which can be easily saved to the attraction database. This database is used by the smartphone for searching information about attraction.

At the first step, a JSON file is generated with a list of selected regions for the attraction information extraction. This list contains region name in the English language, list of region names in terms of information source pages, OSM region identification numbers, local region language, attraction query list, and district query list. Both attraction and district query lists contain subqueries, which were written by using the Overpass API[1] (read-only API that serves up custom selected parts of the OSM map data). Each object in OSM servers has a tag that provides possibilities to identify the object. After the carried out analysis of the OSM tags, the authors concluded that it is reasonable to gather the following inner tags: theaters, museums, religious places, historical places (OSM stores information about castles, monuments, memorials, etc. in this tag), and parks. These tags are the most interesting for the tourists. In case of defining district subquery, an administrative boundary with the specific administrative level in government hierarchy is used. The different countries have different administrative levels for districts, so it is needed to manually specify the level.

For each selected region, the data processing service fetches the region data for constructing the districts. The region data contains information about districts and attractions from OSM servers by using Overpass API. In terms of service, the district within the area of the city is treated as a closed polygon, which can be described by a list of coordinates (see Fig. 3).

The resulting list of coordinates is sorted in the database in the clockwise order. After saving the district information, the service gathers information about region attractions from OSM servers by using the Overpass API. All objects from OSM conceptual data model are represented by a graph with three entities: nodes, ways, and relations. The node defines a specific point with coordinates' one OSM map. The way represents a polyline between two nodes and typically represents roads, rivers, building boundary, etc. The relation is a multi-purpose structure that shows a relation between two OSM objects (e.g., attraction can be treated as set of nodes with ways as boundaries between them). In terms of service, attractions can be represented as nodes with additional information; in case of attraction with "Relation" type, the

---

[1] https://wiki.openstreetmap.org/wiki/Overpass_API.

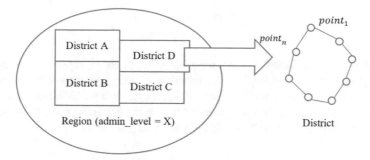

Fig. 3 District representation in the service

conversion process is performed. This process calculates a geographical center of relation and saves it as a node.

Due to the large amount of attraction data, it is reasonable to process it by using separately working threads. The data list is divided accordingly to the number of processor cores, and each part passes to the corresponding thread by service. Each thread processes the geo-data from the OSM servers by fetching the attraction data.

Next, the thread initiates a process of searching for attraction data by using the Wikipedia. In terms of the service, the attraction data contains description, keywords, and images. If the OSM data contains URL link to a Wikipedia page that describes the attraction, the data processing service uses the proposed page; otherwise, the attraction matching service (Sect. 4.2) finds the best suitable information in the Wikipedia.

The following metrics are used to assess the relevance of the proposed attraction page to the attraction:

- Fuzzy string comparison of the proposed attraction name from page and attraction title;
- Search of region via keyword analysis at the proposed attraction page;
- The personality search from the proposed attraction page keywords.

The fuzzy string comparison of the proposed attraction name from the Wikipedia page and the attraction title is done via formula Eq. 1:

$$\delta = \text{ratio}(A_n, P_n) \geq \text{value}, \tag{1}$$

where ratio is a function which calculates the similarity of strings by using the Levenshtein distance and returns a value in the range $[0...100]$; where 0 represents totally different strings and 100 represents identical strings. A_n represents the attraction name and P_n is the attraction name from proposed page given by the attraction matching service. Value is the desired similarity score which can be tweaked empirically (currently, moment is defined as 60).

The criterion for region search by keywords is defined by Eq. 2:

$$\gamma = \begin{cases} \text{true, if any}(K, R_\text{n}) \\ \text{false, otherwise} \end{cases}, \tag{2}$$

where K is a keyword from the proposed attraction page, R_n represents the selected region name and function any(...) is a function of string search in a list. In order to remove abstract articles and attractions, which are not located in the selected region, the metric is looking for the occurrence of the region name in page keywords.

The personality search criterion is defined by Eq. 3:

$$\varepsilon = \begin{cases} \text{true, if any}(K, R_\text{n}) \text{ or any}(K, W_1) \\ \text{false, otherwise} \end{cases}, \tag{3}$$

where W_1 is a specific word for each language connected to the personality search. This metric is only applied to the attraction that has the "monument/memorial" tag in OSM data. This metric is based on the previous one, but includes an additional check aimed at searching for a specific substring in the proposed page keywords. The specific substring is taken from the user-defined string list with personality references (e.g., Russian language—"Персоналии," Finnish language—"Syntyneet," English language—"People," etc.). The user can fill the personality references list. This additional parameter specifies the proposed page as the page, which describes the correlated person with the given attraction.

Equation 4 determines the appropriate attraction page selection. If the proposed attraction page is relevant, the data processing service initiates the extraction process. This process gathers the description and fetches the attraction images. At the moment, the attraction database stores text resources only in English, and multi-language support is not implemented. In case of inaccessibility of the English version of the attraction page in Wikipedia, the service translates the description of the attractions and its keywords, based on categories, by using the Google Translate service.

$$\text{result} = \begin{cases} \delta(A_\text{n}, P_\text{n}) \text{ and } \varepsilon(K, R_\text{n}, W_1), \text{ if type } == \text{"monument/memorial"} \\ \delta(A_\text{n}, P_\text{n}) \text{ and } \gamma(K, R_\text{n}), \quad \text{otherwise} \end{cases} \tag{4}$$

At the end, the data processing service takes all collected attraction data and removes the attraction duplicates. An attraction duplicate is another attraction with the same name and a five-meter distance from the original attraction. This distance was determined practically by analyzing OSM data and attraction locations. Then, the attraction database is filled using the ORM classes. These classes are described in Sect. 4.3. Finally, all objects are saved to the attraction database.

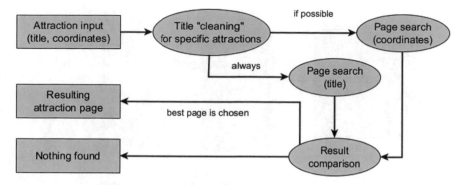

Fig. 4 General process of the attraction page search

4.2   The Attraction Matching Service

The process of an attraction information search in Wikipedia is described in Fig. 4. The search service requires the attraction name, attraction type, and coordinates as input parameters. The title "cleaning" is used if the attraction type is "monument" or "memorial". In this case, the service removes such words as "marshal, academic, memorial, grave," etc. from the attraction title. The user can define the "cleaning" list. This action allows to simplify the further personality identification because page categories or keywords contain the name of personality associated with the attraction.

Then, the basic geo-search by the attraction coordinates and basic search by the attraction name is performed. In case, if the attraction coordinates are missing, only a basic search is performed and coordinates are derived from the Wikipedia. Both types of search take the top five results from the Wikipedia API and identify the most suitable page. The comparison of gathered Wikipedia pages and given attraction name is based on the fuzzy comparison by using the Levenshtein distance. In addition, if the given page contains information about region name in keywords, the service applies additional value to the similarity score. At the end, the similarity results of geographic and basic search are compared to each other, and the attraction with the highest score is selected.

4.3   The Attraction Delivery Service

The attraction delivery service stores attraction databases for each region and provides them to the tourists. Databases are created in SQLite format. This action allows users with smartphones to download region database with all attraction information and use it offline. To reduce the weight, all database instances are packaged in zip archives. The database structure is defined in Fig. 5. The main ORM classes are *Attraction*, *Keyword*, *Media,* and *District*. The *Attraction* class contains information

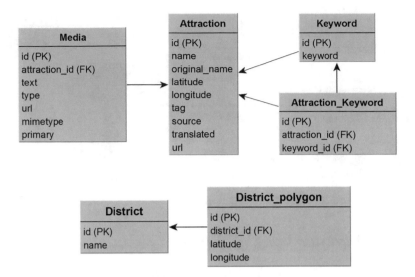

Fig. 5 Database scheme for ORM objects

about attraction name (English and local language), coordinates (latitude and longitude), tag (such as "museum," "memorial," etc.), source of detailed information (Wikipedia, Wikivoyage, etc.), attraction page URL, and other identifiers and flags (e.g., if the translation of the information has been initiated). The class *Keyword* contains the list of keywords for the attraction. At the moment, the attraction keywords are populated from the Wikipedia attraction page categories. The class *Media* contains the attraction-related content. In this class, the text description and audio/video information are stored. The class *District* contains information about districts of selected region with district name and coordinates polygon.

4.4 The Attraction Managing Service

The attraction managing service is aimed to manipulate the database developed by the data processing service. The service has a Web-based administration panel, which makes it possible to add, remove, and update information about attractions for each region. It supports the possibility to include audio/video media files about attractions and manage the textual information. The general view of the service is presented in Fig. 6.

The service administrator is able to view the current region with statistics about attraction, keywords, and media files included in the database, edit information (download files, edit the region content, or remove region entry from the database), and create the new region. If the administrator decides to create a new region, he/she needs to specify the region name, location and provide a title image for the region.

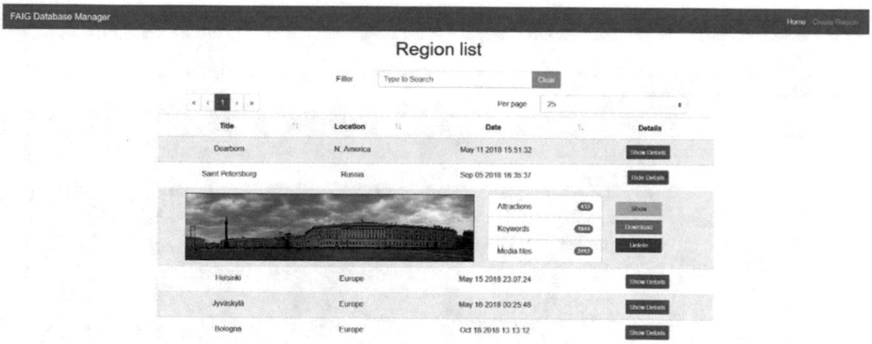

Fig. 6 Attraction managing service: the main page

At the region details page (Fig. 7), all region attractions are specified. Due to the large number of attractions, the possibility of filtering and pagination was introduced. All elements are grouped into the table. Each row contains basic information about the attraction including title and coordinates as well as "Show details" and "Action" buttons. The "Show details" button displays the additional information: original attraction title, tags, information source type, Web page URL, keywords, and media files such as text description and audio/video files. By using the "Action" button, the

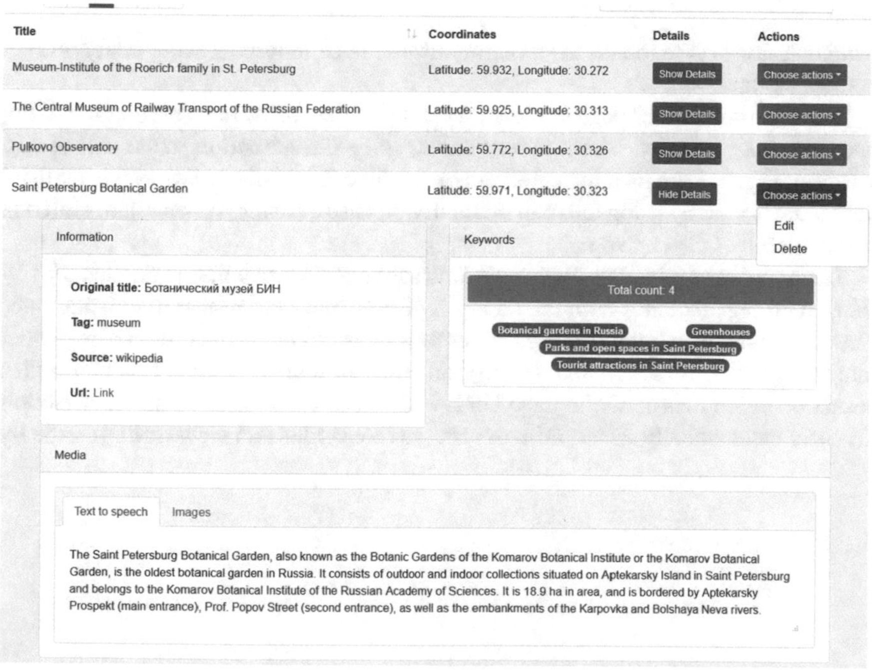

Fig. 7 Attraction managing service: the region details page

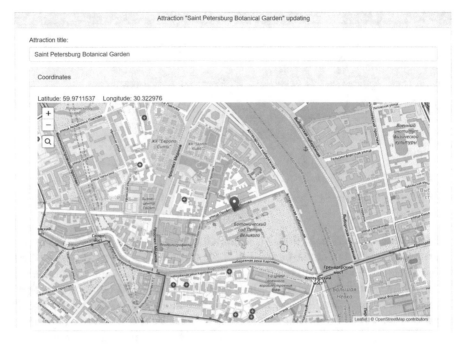

Fig. 8 Attraction managing service: the coordinate selection

administrator is able to edit the existing attraction or remove it completely with all associated data.

The creation/editing page is aimed to create a new attraction or edit an existing one. On this page, the administrator can specify the attraction name and choose location (Fig. 8). The location selection on a map can be done in two ways: by using the search box or by using the map manually. In case of using the searchbox, all data is provided by OSM servers.

The user can specify tags for selected attraction and work with media files (Fig. 9). Work with tags includes both the creation of new ones and re-use of existing ones. The attraction managing service provides tags to user. The media files can be divided into five groups: attraction text description, attraction images, attraction audio files, attraction video files and Web page URL with specific information about attraction. The user can manually create, edit, or remove media files and view results on the fly.

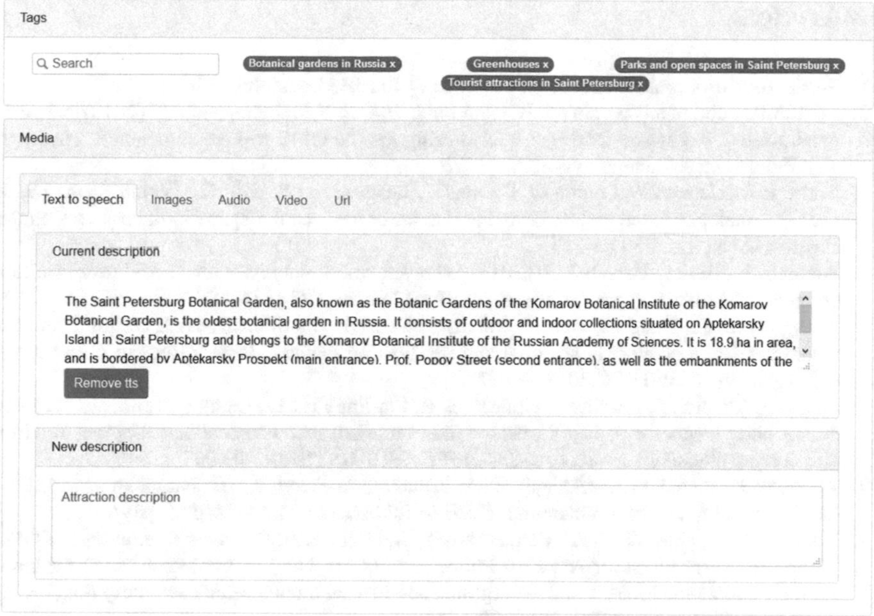

Fig. 9 Attraction managing service: the tags and media interaction

5 Conclusion

The paper presents an approach to context-driven tourist trip planning system development. The core element of the system is the attraction database. The database is created based on information acquired from OSM and Wikipedia by the data processing and attraction matching services. The developed attraction managing service is used by tourism experts that can add information about attractions to the database through the HTML-based client. Attraction delivery service extracts information from the attraction database that can be interesting to the tourist based on his/her context. At the moment, the presented system is designed for English language, and it stores only one text fragment for every attraction. In the future, the authors plan to implement multi-lingual support and possibility to store different text descriptions for every attraction for presenting these to tourists based on the current situation, their preferences, and trip plan.

Acknowledgements The related research and reference model of the tourist trip planning system have been carried out in the scope of Grant № 18-37-00337 of the Russian Foundation for Basic Research. The attraction managing and delivery services have been developed in the scope of Grant № 17-29-03284 of the Russian Foundation for Basic Research. The overall scheme of the on-board dynamic tour support system has been developed in the scope of Ford University Research Program. Implementation of the tourist trip planning system has been done in the scope of Project № 618268 supported by ITMO University.

References

1. World Tourism Organization (UNWTO) (2017) Tourism highlights, p 16
2. Logesh R, Subramaniyaswamy V, Vijayakumar V, Gao X-Z, Indragandhi V (2018) A hybrid quantum-induced swarm intelligence clustering for the urban trip recommendation in smart city. Future Gener Comput Syst 83:653–673
3. Bartie P, Mackaness W, Lemon O, Dalmas T, Janarthanam S, Hill RL, Dickinson A, Liu X (2018) A dialogue based mobile virtual assistant for tourists: the SpaceBook Project. Comput Environ Urban Syst 67:110–123
4. Amoretti M, Belli L, Zanichelli F (2017) UTravel: smart mobility with a novel user profiling and recommendation approach. Pervasive Mob Comput 38(2):474–489
5. Zhang X, Yu L, Wang M, Gao W (2018) FM-based: algorithm research on rural tourism recommendation combining seasonal and distribution features. Pattern Recogn Lett. https://doi.org/10.1016/J.PATREC.2018.12.022
6. Santos F, Almeida A, Martins C, Gonçalves R, Martins J (2017) Using POI functionality and accessibility levels for delivering personalized tourism recommendations. Comput Environ Urban Syst. https://doi.org/10.1016/J.COMPENVURBSYS.2017.08.007
7. Al-Rayes K, Sevkli A, Al-Moaiqel H, Al-Ajlan H, Al-Salem K, Al- Fantoukh N (2011) A mobile tourist guide for trip planning. IEEE Multidiscip Eng Educ Mag 6(4):1–6
8. Vdovenko A, Lukovnikova A, Marchenkov S, Sidorcheva N, Polyakov S, Korzun D (2012) World around me client for Windows Phone devices. In: The 11th conference of Open Innovations Association FRUCT and 1st regional seminar on mobile healthcare, early diagnostics and fitness, St-Petersburg, Russia, pp 206–208
9. Garcia O, Alonso R, Guevara F, Sancho D, Sánchez M, Bajo J (2011) ARTIZT: applying ambient intelligence to a museum guide scenario. In: Ambient intelligence - software and applications. Springer, Berlin, Heidelberg, pp 173–180
10. Smirnov A, Kashevnik A, Ponomarev A (2017) Context-based infomobility system for cultural heritage recommendation: Tourist Assistant—TAIS. Pers Ubiquit Comput 21(2):297–311
11. Mikhailov S, Kashevnik A (2018) Smartphone-based tourist trip planning system: a context-based approach to offline attraction recommendation. In: Ronzhin AL, Shishlakov VF (ed) 13th international scientific-technical conference on electromechanics and robotics "Zavalishin's Readings" (ER(ZR)-2018), MATEC web of conferences, vol 161. EDP Sciences, p 03026
12. Shilov N, Kashevnik A, Mikhailov S (2018) Context-aware generation of personalized audio tours: approach and evaluation. In: Karpov A, Jokisch O, Potapova R (eds) 20th international conference on speech and computer (SPECOM 2018). Lecture notes in artificial intelligence, vol 11096. Springer International Publishing, Switzerland, pp 615–624

Regional Geoinformation Modeling of Ground Access to the Forest Fires in Russia

Ekaterina Podolskaia, Konstantin Kovganko and Dmitriy Ershov

Abstract Every year forest fires in Russia, arising on large areas, cause signifi-cant damage to the forests, infrastructure and livelihoods of local population. Paper presents an overview of current forest firefighting situation in Russia. Interdisci-plinary research combining the specifics of road infrastructure and Russian forest firefighting regulations was conducted to improve ground forest firefighting opera-tions at the regional level and to propose a geoinformation solution. We described a workflow to create the shortest access routes (using Dijkstra algorithm, ArcGIS Desktop Network Analyst) from the fire stations to the forest fires taking into account traveled distance, time and elevation of transport network (public roads and forest glades). Data of access routes created for 2002–2017 forest fire hazardous seasons have been processed and analyzed. Developed service "Access routes" creates the daily-updated cartographic products. Irkutsk region in Siberia, Russia was chosen as a study area.

Keywords Logistics · Transport model · ArcGIS · Network analyst · Forest fire

1 Introduction

In this paper, we consider the thematic GIS transport modeling in application to the forest firefighting in Russia, taking into account publicly available official documents on the timely ground access to the forest fires from the fire stations ("fire-chemical stations" in Russian forest terminology). Prompt access of fire brigade and delivery of special equipment to the place where a forest fire is detected is one of the most challenging topics in the forestry industry. Systems that can solve described task have a certain need for the Russian economy, especially taking into account large spatial extent of the country and its regional needs.

E. Podolskaia (✉) · K. Kovganko · D. Ershov
Center for Forest Ecology and Productivity of the Russian Academy of Sciences (CEPF RAS),
Profsoyuznaya st. 84/32 bldg. 14, 117997 Moscow, Russian Federation
e-mail: podols_kate@mail.ru

© Springer Nature Switzerland AG 2020
V. Popovich et al. (eds.), *Information Fusion and Intelligent*
Geographic Information Systems, Advances in Geographic Information Science,
https://doi.org/10.1007/978-3-030-31608-2_11

Our goal is to propose a GIS-workflow to create the shortest access routes. There are two tasks to complete it: (1) to create and to test transport model to assess the forest fires from the fire stations (stage of geoinformation modeling), and (2) to propose a GIS-service for the fire hazardous season's daily monitoring (stage of development). Practical implementation of the transportation task in GIS environment is the subject of our study. *A forest fire* is a temperature anomaly detected by satellite Terra/Aqua on the platform MODIS and localized in points. *An access route* is a created polyline of shortest (according to the Dijkstra algorithm) distance. *A forest hazardous season* is a time period when a forest fire can occur. First results were delivered and discussed at the Conferences "Forests of Russia" [1] and "Scientific basis for sustainable forest management" [2].

2 Research Context

We cover features of transportation systems and forest legislation in Russia, as well as role and spatial distribution of fire stations at regional level.

2.1 Researches on Transport Systems

Research on transportation is conducted in the areas of mathematical modeling and algorithms such as shortest path, vehicle routing and facility location [3–5], and in the geoinformation domain [6–9]. Term "transportation" is usually related to the collection of "transportation task," "finding a path," "optimal path problem," "travelling salesman problem (TSP)," and "vehicle routing problem (VRP)" widely used in the scientific and technical literature.

Undertaken analysis of published papers shows a good level of topic's development at large-scale mapping and organization of transportation within populated areas (settlements like cities or towns), for instance, in different parts of the world [10, 11]. One of the examples of Russian and foreign practice is an information managing system for a city or a metropolis [12, 13]. Smaller number of publications were made where two areas, namely the transportation task (*"transport"*) and *"logistics"* [14–16] and forest fire safety (*"fires"*) [17, 18] are joined, in particular at the regional scale, away from the populated areas with their high level of infrastructure's development.

Optimization GIS-approaches are described in the number of papers. Authors use the definition of "optimal transportation plan" and propose some interactive cartographic systems [12, 14, 19], they discuss project management and specifics of decision making in transportation [20, 21]. Transport accessibility is the subject of research for some transportation studies like [22].

Nowadays tools of different proprietary status are being widely used. Modern GIS in fact has a multifunctional toolkit for solving the transportation problem in a significant number of its applications. For instance, ArcGIS Network Analyst extension with its Dijkstra algorithm is implemented to the city road system or to the construction of roads network [13, 23, 24]. Network Analysis and Geoprocessing Services (or Samples) are being published and available for the users, there are also commercial stand-alone software products.

Terminology of transportation is also evolving, and its standardization is an ongoing process. To summarize, a road network is a set of ordered nodes (crossroads, junctions, stations, etc.) and edges (special arcs of transport graph which connects the conditional center of the area with the node of the transport network); a weight is assigned to each road section or segment. Following concepts are important in application to the "time" and "distance" definitions such as: value of time, scenario, traffic supply, and traffic zones.

Analyzing the publications, we need to highlight that the researches were active in recent years, and obviously this fact is related to the development of GIS and its diverse applications. Travel-time impact and factors, as well as the procedures to estimate time, were considered in the book [21]. Time is a parameter which plays a key role in any optimization process [25, 26]. Similar goal and relevant to the present paper experience are described in the papers [6, 27–29]. Overall, transportation task remains multiparameter [30] and requires the knowledge of thematic area (specifics of country or region) like it is shown in the papers, to mention a few [31, 32], and therefore it is an actual research problem.

2.2   Forest Fires Situation and Forest Legislation in Russia

There are several official Russian organizations responsible for the forest firefighting: Aerial Forest Protection Service, Federal Forestry Agency, and Ministry of Emergency Situations. These organizations publish forestry regulations at the federal and regional levels.

The problem of forest firefighting (including their monitoring, assessment, and access) remains important in Russia [33, 34]. Efforts have been made by scientific and technical communities: There is a short- and long-term prognosis of how the forest firefighting situation can be evaluated and improved [34], specific methods to establish an effective forest firefighting system [18] are proposed, fire forests modes and their classification [35] are described. Necessity and importance to organize timely response to an emergency, rapid mobilization and delivery of material and human resources have been emphasized in many papers in Russia and abroad, for instance [18]. Underdeveloped public roads and forest glades in the remote regions is an additional challenge to be considered when accessing the forest fires.

There are some valid official documents that set up the forest firefighting rules, namely [36–40]. Generally, forest fire hazardous season in Russia lasts more than half a year, starting from the first of April and ending on the first of November, depending on the region and weather conditions. Several types of forest zones' control are identified, ground and aviation protection zones are among them [38, 39]. Access has to be organized within three hours from the moment of forest fire's detection in the ground protection zone [36]. During this period, special measures are established in the forest zones, for example, the restricted public access and daily monitoring. Five classes of natural fire danger are known, they differ from "no danger" to "emergency" statuses [40] to characterize the probability of fires occurrence's in relation to the meteorological conditions.

2.3   Fire Stations for the Forest FireFighting

According to the document [37], a *fire station* is a forest department which is organized in accordance with the regional forest firefighting plans; spatial distribution of stations is analyzed and discussed when fire hazardous season is being prepared, practically on the annual basis. There is a procedure to manage a forest fire when it has been detected. Forest manager or ranger, having received a report on a fire, is responsible to take actions and to manage the fire station's fire team or brigade. Fire brigade should arrive at the forest fire and start the extinguishing works. Stations' stuff initiates, undertakes, and completes the extinguishing works; stations are being organized "on-site" and can share their material resources with other neighborhood stations if needed.

Spatial distribution of stations in the region depends on the level of natural forest fire danger [40]. According to the official regulation [37], there are three types of fire stations: I, II, and III. Stations of first type are organized for the fire hazardous season only, they have technical means to manage up to two forest fires per day. Stations of II type are organized seasonally or permanently to manage up to four forest fires daily. And, finally the III type is located in the regions with high natural forest fire danger on the permanent basis and has the most powerful firefighting transport and equipment. In many cases, fire stations in the Russian regions are located in the settlements, mainly villages. Thus, a fire station plays an important role in the forest firefighting.

3   Preparation of Regional Transport Network

Let us consider the datasets and give a description of the study area, the attributes to calculate for the regional transport network. Original digital data used in the present work have been taken from the public sources and have been kindly provided by our partners, mainly by the Irkutsk regional hub. We use scale 1:200,000 as reference because it has sufficient level of map generalization for the regional projects, that is why it was our reference to produce more generalized layouts (1:500,000, 1:1,000,000 and others).

Study area (Irkutsk region) is located in the Central Siberia, a part of Siberian Federal district, Russia. This area of interest was chosen because of an intensive annual forest fire activity. It is important to note that data on the forest glades within the test area are considered as an element of the transport network (about 20% of the total roads length, according to our calculation), assuming that firefighting trucks can move along them depending on glades' actual condition.

Any route optimization technique would consider the distance and time or speed. For the present research, we have used the factors of *Speed, Time,* and *Elevation*.

Speed: In order to establish the speed limits for the fire trucks in remote areas (mainly covered by unpaved roads), the authors examined the existing Russian rules. Roads classification was prepared with the Speed (km per hour) attribute. This classification depending on the road's type was used later on to calculate traveled time of road segment.

Elevation: A certain number of Russian regions with the annual forest fire activity are very hilly, especially in Siberia. It becomes crucial when any ground displacement (fire trucks, stuff, and heavy equipment) to a forest fire has to be planned. Territory coverage and spatial resolution were taken into account to choice the Global Elevation Model ETopo2 [41]. Such model can be of different type and source, for example, created from the locally or regionally available data, or any global model, including freely available on the Internet. Elevation as an essential parameter for the regional transportation (in the present research—because of large extent of test area) has been already used as a parameter for the transport optimization in the number of published works [9, 42].

Time: In our approach, "Time 3D" (time to travel three-dimensional length of each polyline) in minutes is calculated with "Length 3D" (three-dimensional length of each polyline) in kilometers.

3D length of a road segment is either equal to, or greater than 2D length. This is a useful "adjustment" to the polyline's accuracy to calculate the entire access route. Slope can be either "positive" or "negative," depending on where we start movement. In both cases ("positive" or "negative" slope), we decrease the speed value.

For polyline datasets (public roads and forest glades), we added the attributes of *Speed* and *Time,* both calculated with *Elevation*. So, we defined the content and data attributes for the regional transport network.

4 Proposed Workflow, Tests, and Analysis

Study design includes several steps (Fig. 1): from collecting and preparing the road data (public roads and forest glades) to a GIS view. We developed a model (including road data, fire stations and forest fires) to create the access routes and to estimate traveled distance and time with elevation values. By running the model in the automatic mode, we create the access routes and, finally, put their graphical and attributive representation into the GIS environment.

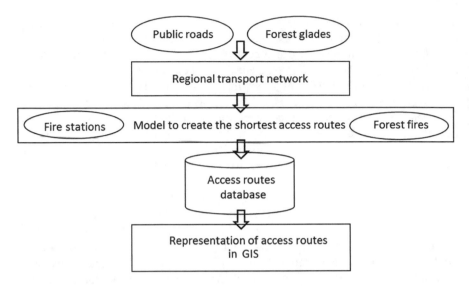

Fig. 1 Study design

4.1 Model to Create Access Routes and to Estimate Their Characteristics

Polyline datasets with roads and glades were imported into the file geodatabase to create a transport network. Network's description with its properties and attributes is given in Table 1.

To automate the process of routes creation, we have developed a model in the Model Builder visual programming environment. In the model where forest fires and locations of fire stations are used (Fig. 1), the first step is to run solve standard procedure. Then, resulting network completes with new attributes such as "Number of Forest Fire," "Number of Fire Station," and "Date of event."

At the end of described workflow, we put the access routes to the relational database management system (RDBMS MySQL is established) and export them to a template document (layout to print). Access route is being created when any new forest fire occur, for the forest fire hazardous season usually on the daily basis. As a result of modeling, one route per forest fire is created. Data on the forest fires from the previous years and from ongoing forest fire hazardous season can be used as input. So, we developed a forest firefighting transport model of ground access. The model is used to generate access routes, to accumulate and to analyze their statistics.

Table 1 Description of Irkutsk regional transport network

Properties	Attributes
Name: Transport routes Type: Geodatabase-based Network dataset Edge sources: Transport routes Connectivity: Edge connectivity: Transport routes (any vertex) Elevation model: Z coordinate values	**Length**: Usage type: Cost Data type: Double Units type: Kilometers Source attribute evaluators: Transport routes (from-to): Field Language: VBScript Expression: [Shape] Roads Irkutsk (to-from): Field Language: VBScript Expression: [Shape] **Time3D Speed**: Usage type: Cost Data type: Double Units type: Minutes Use by default: True Source attribute evaluators: Roads Irkutsk (from-to): Field Language: VBScript Expression: [Time Min] Transport routes (to-from): Field Language: VBScript Expression: [Time Min]

4.2 Key Results of the Model's Testing

To test the model, we have processed the data on the forest fires retrospectively from 2002 (when the federal monitoring system has been launched and forest fires data became available on the regular basis) to 2017 for the Irkutsk region. Calculations and analysis of results for the shortest access routes were carried out for the forest fires registered within ground protection and forest aviation zones for every fire forest hazardous season. Important to note that dataset with forest protection zones is dated of 2017. *Number* of created routes and their *length* have been analyzed for the three time zones (one, two, and three hours of access).

Number of routes: The minimum number of routes (236) is built for 2004, the maximum (3297) is for 2003 which is due to the random number of detected forest fires. Number of days with forest fires occurred during the fire hazardous season varies from 170 to 213 days.

Most of the forest fires (2661 or 98.5%) detected within the ground protection zone are accessible within three hours. This result indirectly confirms the correctness of model's settings, focusing on the fact that the ground protection zone was spatially designed under "three hours maximum" condition of ground access. 1754 or 68.2% of forest fires within the ground zone could have one-hour access limit, 744 or 24.3%—two hours. So, majority of forest fires would be accessible by ground technical means.

We observe the same tendency for the first two months (April–May) of fire hazardous season 2018 with 39 forest fires detected. The access time varies from 13 to 185 min; 30 fires are accessible within one hour, 4 fires—one-two hours; 4 fires—two-three hours, and 1 fire event—more than three hours.

Length of routes: For every year, we have calculated the total length of routes within two protection zones, the average relative value of routes within three-hour availability is about 50%. Accessibility of less than three hours characterized the relative length of routes (in %) created for 2002–2011, and then in 2012–2017 accessibility of more than three hours is dominated.

By analyzing these two numerical parameters (number and length) of created routes, we have shown graphically (on the map and charts) and statistically (by calculations) that ground protection zone in the Irkutsk region could be possibly extended but these statistics have to be validated before any action.

4.3 GIS-Service "Access Routes"

Described methodology led us to create a solution to present the visual analytical results in the form of cartographic layouts to support the specialists (employees of fire stations) in the situation of making logistical and transportation decisions.

Developed GIS-service named "Access routes" is aimed to be used at the regional scale. We have a cartographic interface displaying created shortest access routes, locations of forest fires, fire stations with their attributes and reference topographic data with the legend. Regional datasets are shown at three scales (1:200,000, 1:1,000,000, and 1:7,000,000). Russian is the language of the user's interface.

Proposed workflow allows to create a graphical output file with the access routes and forest fires localization automatically on the daily basis, file can be retrieved from the GIS environment by the users. Set of layers could be customized with the static or dynamic Services (Map, Geoprocessing, etc.) depending on the needs of local users.

5 Conclusions

We have presented an example of geoinformation solution to solve the transportation task at the regional scale taking into account the forest fire network's parameters (traveled time and length, both with elevation values). Technical value of the described approach is the combination of developed workflow on how to create the shortest routes with the multiple data sources (forest firefighting infrastructure and geographical data) into one GIS interface available for the regional needs in Russia.

We have processed the data and produced the archive of access routes for the forest fires detected in 2002–2017 (totally 16 years). Analysis of assess routes data allowed us to give the recommendations on how to distinguish the technical means of access to the forest fires: by forest fire trucks (ground access within three hours which is required by Russian forest legislation), and then by firefighting airplanes. Daily monitoring for the ongoing fire hazardous season and for the years to come has been established in the GIS form. Thus, the tasks described in the introduction were completed. Important result in our workflow is that by analyzing the archive of created access routes we can indirectly estimate spatial correctness of actual protection zones.

With every new forest fire hazardous season, we will have more information about the spatial distribution and length of the routes. Developed GIS-service will help in the preparation and planning of forest fire hazardous season and in the cooperation of efforts between ground and aviation forest firefighting teams.

The project is ongoing, and its next stage is dedicated to the validation. It is necessary to highlight that we work with the regional Irkutsk dispatching hub in order to obtain the real tracking datasets, to analyze them and to calibrate the parameters of the transport model described in this paper. Modeled access routes would be compared to the real routes of trucks obtained in collaboration with the regional authority.

Acknowledgements This work was supported by the state contract "Development of methodological approaches to remote monitoring of resource potential and ecological state of forest ecosystems" (topic No. 0110-2017-0001).

References

1. Podolskaia ES, Kovganko KA, Ershov DV, Plotnikova AS (2017) Creation of geoinformation model to plan an optimal route for ground access to the forest fires (in Russian). In: Gedio VM (ed) Proceedings of II international scientific and technical conference "forests of Russia: politics, industry, science and education". Saint-Petersburg State Forest Technical University, pp 200–202
2. Podolskaia ES, Ershov DV, Kovganko KA (2018) Modern possibilities of GIS to model the ground access to the forest fires (in Russian). In: III All-Russian scientific conference with international participation "scientific basis for sustainable forest management". http://cepl.rssi.ru/confs/forest_management_2018/files/forest_man_program3_29102018.pdf
3. Taner F, Galic A, Caric T (2012) Solving practical vehicle routing problem with time windows using metaheuristic algorithms. Promet-Traffic Transp 24(4):343–351
4. Vorobiev AE, Titov AY, Gavrilin VA, Menshutin AY, Bakhirev IA (2015) Transport model of Moscow region (in Russian). Methods of supercomputer modeling based on Intercosmos. Institute of Space Researches, Russian Academy of Science. http://www.iki.rssi.ru/seminar/20151111719/
5. Caha J, Dvorsky J (2015) Optimal path problem with possibilistic weights. In: Geoinformatics for transportation. Springer, Berlin. https://www.springerprofessional.de/optimal-path-problem-with-possibilistic-weights/2278722
6. Shankar H, Mani G, Pandey K (2014) GIS based solution of multi-depot capacitated vehicle routing problem with time window using tabu search. Int J Traffic Transp Eng 3(2):83–100
7. Ford AC, Barr SL, Dawson RJ, James Ph (2015) Transport accessibility analysis using GIS: accessing sustainable transport in London. ISPRS Int J Geo-Inf 4:24–149
8. Abousaeidi M, Fauzi R, Muhamad R (2016) Geographic information system (GIS) modeling approach to determine the fastest delivery routes. Saudi J Biol Sci 23:555–564

9. Demidov VK, Pivovar AI, Kravchenko EG (2017) Finding a path and its visibility based on digital elevation model in open sources GIS (in Russian). In: Geoinformatics-2017, EAGE
10. Kumar P, Singh V, Reddy D (2005) Advanced traveler information system for Hyderabad city. IEEE Trans Intell Transp Syst 6(1):26–37
11. Idhoko KE, Aguba W, Emefeke U, Nwanguma C (2015) Development of a geographic information systems road network database for emergency response: a case study of Oyo-Town, Oyo State, Nigeria. Int J Eng Sci Invent ISSN 4(12):34–42
12. Liu S, Zhu X (2004) Accessibility analyst: an integrated GIS tool for accessibility analysis in urban transportation planning. Environ Plan Plan Des 31:105–124
13. Forkuo EK, Quaye-Ballard JA (2013) GIS-based fire emergency response system. Int J Remote Sens GIS 2(1):32–40
14. Sokolov AP, YY Gerasimov (2009) Geoinformation system for solving the optimization problem of round timber logistics (in Russian). IVUZ For J N3:78–85
15. Tavares G, Zsigraiova Z, Semiao V, Carvalho MG (2009) Optimization of MSW collection routes for minimum fuel consumption using 3D GIS modelling. Waste Manage 29:1176–1185
16. Grigolato S, Mologni O, Cavalli R (2017) GIS applications in forest operations and road network planning: an overview over the last two decades. Croat J Eng 38(2):175–186
17. Dunna CJ, Thompson MP, Calkin DE (2017) A framework for developing safe and effective large-fire response in a new fire management paradigm. For Ecol Manage 404:184–196
18. Kotelnikov RV, Korshunov NA, Giryaev NM (2017) Objectives of decision making in protecting forests from fires. Main priorities on development of informational support (in Russian). Sibirskij Lesnoj Zurnal (Siberian J Sci) 5:18–24
19. Alazab A, Venkatraman S, Abawajy J, Alazab M (2011) An optimal transportation routing approach using GIS-based dynamic traffic flows. In: 3rd International conference on information and financial engineering IPEDR, vol 12. IACSIT Press, pp 172–178
20. Kurganov VM, Dorofeev AN (2015) Information systems for road transportation (in Russian). Mir transporta (World Transp Transp) 13(3):165–171
21. Sinha KC, Labi S (2017) Transportation decision making principles of project evaluation and programming. Wiley, Hoboken, 537 p
22. Handy SL, Clifton KJ (2001) Evaluating neighborhood accessibility: possibilities and practicalities. J Transp Stat 67–78
23. Ni K, Zhang Y-T, Ma Y-P (2014) Shortest path analysis based on Dijkstra's algorithm in emergency response system. Telkomnika Indonesian J Electr Eng 12(5):3476–3482
24. Parsakhoo A, Jajouzadeh M (2016) Determining an optimal path for forest road construction using Dijkstra's algorithm. J For Sci 62(6):264–268
25. Alekseev SP (2015) Reducing arrival time of the fire brigade to the fire places (in Russian). Internet J Technol Technosphere Secur 4(62). http://agps-2006.narod.ru/ttb/2015-4/32-04-15.ttb.pdf
26. Jawad FF, Shabana BT, El-Bakry HM (2016) Reducing waiting time for transportation using GIS. Int J Adv Res Comput Sci Technol (IJARCST) 4(1):62–71
27. Akay AE, Wing MG, Sivrikaya F, Sakar DA (2012) GIS-based decision support system for determining the shortest and safest route to forest fires: a case study in Mediterranean Region of Turkey. Environ Monit Assess 184:1391–1407
28. Akay A, Aziz B (2015) GIS-based forest road network model for forest protection purposes. In: 38th Annual COFE meeting. Engineering solutions for non-industrial private forest operations, pp 266–281
29. Akay AE, Suslu HE (2017) Developing GIS based decision support system for planning transportation of forest products. J Innov Sci Eng 1(1):6–16
30. Kobrin OV, Tishayev IV, Zatserkovnyi VI (2017) Modeling of parameters and optimization of transport networks with the help of GIS (in Russian). In: 16th International conference on geoinformatics. Theoretical and applied aspects, EAGE. http://www.earthdoc.org/publication/publicationdetails/?publication=89565
31. Kovalev KE, Galkina YE, Vasiliev AB (2016) Transportation task on railways with operational and economic criteria (in Russian). In: "Transport of Russia: challenges and perspectives—2016" Proceedings of the international scientific and practical conference, vol 2, pp 283–287

32. Korovyakovskiy EK, Badetskiy AP (2016) Vehicles route choosing under uncertain conditions (in Russian). In: "Transport of Russia: challenges and perspectives—2016" Proceedings of the international scientific and practical conference, vol 2, pp 280–283
33. Goldammer JG, Sukhinin A, Csiszar I (2003) Current fire situation in the Russian Federation: implications for enhancing international and regional cooperation in the UN Framework and the global programs on fire monitoring and assessment. In: GFMC contribution to the international workshop "new approaches to forest protection and fire management at an ecosystem level", Khabarovsk, Russian Federation, pp 1–24
34. Loupian EA, Mazurov AA, Flitman EV, Ershov DV, Korovin GN, Novik VP, Abushenko NA, Altyntsev DA, Koshelev VV, Tashchilin SA, Tatarnikov AV, Csiszar I, Sukhinin AI, Ponomarev EI, Afonin SV, Belov VV, Matvienko GG, Loboda T (2006) Satellite monitoring of forest fires in Russia at federal and regional levels (in Russian). In: Mitigation and adaptation strategies for global change, vol 11. Springer, Berlin, pp 113–145
35. Volokitina AV, Sofronova TM, Korets MA (2016) Regional scales of fire danger rating in the forest: improved technique (in Russian). Sibirskij Lesnoj Zurnal (Siberian J For Sci) 2:52–61
36. Methodical recommendations on the use of forces and technical means to extinguish the forest fires (Emercom, from 16.07.2014 N 2-4-87-9-18). http://legalacts.ru/doc/metodicheskie-rekomendatsii-po-primeneniiu-sil-i-sredstv-dlja-tushenija/ (in Russian)
37. Regulations on fire stations, dated of 19/12/1997. N 167. http://docs.cntd.ru/document/58817250 (in Russian)
38. Order Federal Forestry Agency N 65 from 16.02.2017. http://docs.cntd.ru/document/456071028 (in Russian)
39. Order from the Forest Ministry of Irkutsk region, N 52 from 21.06.2017. http://docs.cntd.ru/document/450255889 (in Russian)
40. Classification of natural forest fire danger depending on weather conditions, order N 287. Federal Forestry Agency, 2011. https://rg.ru/2011/08/24/pojari-dok.html. (in Russian)
41. https://www.ngdc.noaa.gov/mgg/global/etopo2.html
42. Verma PA, Kumar K, Saran S (2017) Road network impedance factor modelling based on slope and curvature of the road. Int J Adv Remote Sens GIS 6(1):2274–2280

Concept of Intelligent Decision-Making Support System for City Environment Management

Elena Batunova, Tatiana Popovich, Oksana Smirnova and Sergey Truhachev

Abstract Through recent years, public's requirements for comfort and safety of cities' infrastructure have considerably risen. Naturally, cities' governments have striven to satisfy demands of their citizens. An innovative concept of intelligent decision-making support system for city environment management, presented in this paper, is oriented towards integration of modern GIS-technologies, information technologies and decision-making tools into city planning and management processes. This concept was developed as a part of international collaborative project CRISALIDE under the international program ERA.Net RUS PLUS. In this paper, we discuss major problems associated with transition of a city to a smart city model and various approaches to realisation of this model. Within this paper, we also provide estimation of prospects for further development of smart city concepts in Russia and overseas.

Keywords Smart city · Intelligent GIS · Decision-making support system · City management

1 Introduction

The dynamic of life in modern cities grows from year to year. This can primarily be attributed to high pace and scale of economic development of cities, to scientific and technological progress in various spheres including information technologies, to implementation of innovative technologies in industry, and to continuous population growth. Modern dynamics of development of megacities influences all spheres of life of modern citizens: their professional and personal interests and daily needs.

E. Batunova · S. Truhachev
Southern Urban Planning Center, Rostov-on-Don, Russia

T. Popovich · O. Smirnova (✉)
SPIIRAS Hi Tech Research and Development Office Ltd, St. Petersburg, Russia
e-mail: sov@oogis.ru

T. Popovich
e-mail: t.popovich@oogis.ru

© Springer Nature Switzerland AG 2020
V. Popovich et al. (eds.), *Information Fusion and Intelligent Geographic Information Systems*, Advances in Geographic Information Science,
https://doi.org/10.1007/978-3-030-31608-2_12

167

Concepts that are related to development of modern city infrastructure and that implement modern information technologies are called smart cities. Basic components of such concept are energy efficiency, environmental friendliness, safety, accessibility of various services, electronic services, etc.

It is worth mentioning that interests and expectations of citizens and that of authorities and management companies are very different. For the citizens, the issue of most importance is convenience of services and payments, including housing and communal services, public transport, parking, medical and government services. For the city government, the main priorities are public security, effective management of emergency and operational services, city traffic management, income and expenditure planning, ecological monitoring, transparency of educational and medical services and establishment of "single window" regime for citizens and companies accessing public services.

Management and utility-providing organisations have other priorities: automated collection of data from metering devices, invoicing and payment control, prediction of resource consumption, energy-saving technologies, weather-oriented engineering systems' operation, provision of automated and manual system deactivation in case of natural and technological disasters, repair planning and debureaucratization of interaction with city government.

However, it should be noted that in order to achieve sustainable development of city environment, to provide ecological security, to implement social innovations along with information and communication technologies, city environment management with application of innovative technologies should be oriented primarily on national and local authorities, companies and organisations and also on citizens themselves. The basis of such system is modern intelligent methods and geoinformation systems.

This paper outlines the design of the concept of intelligent decision-making support system for city environment for the CRISALIDE project under the international program ERA.Net RUS PLUS. The goal of this concept is to create conditions for effective decision making for city management as entire economic and social system. Application of modern information technologies for strategic and operational management of the city environment could bring significant economic benefits.

2   Market Analysis and Problem Statement

According to Bank of America Merrill Lynch's report, 55% of the world's population currently resides in cities and this number is expected to grow up to 70% by 2050 [1]. It is well known that cities have clearly become the engines of global economic growth all over the world. However, more than 80% of the world's cities were found to show signs of fragility [1]. There are similar critical issues in practically every city that are yet to be fully solved [1]:

- poor governance and weak institutions, which are perceived to be the major impediment to city's prosperity;
- inadequate infrastructure;
- rising inequality among cities (75% of cities are worse off than 20 years ago);
- housing;
- crime;
- environmental challenges;
- new and pervasive risks (cybersecurity, terrorism, securitisation, disease and pandemics, etc.).

At the meantime, smart cities seek to tackle these issues using twenty-first century disruptive technologies. These include [1]: ubiquitous broadband coverage (84% globally); nextgen infrastructure; the Internet of Things; Big Data; the Cloud (secure, open platform); and artificial intelligence (AI) (predictive insights, anticipatory actions).

The most popular directions, in which smart cities currently take their development, are the following [1]:

(1) smart infrastructure;
(2) smart buildings;
(3) smart homes;
(4) smart safety and security;
(5) smart energy;
(6) smart mobility.

It should also be noted that one of the most advanced and popular modern technologies that are fully capable to address demographic, economic, environmental, infrastructural and social issues in smart cities is geoinformation systems (GIS).

Technavio's market research analysts have predicted that the GIS market will register a CAGR of more than 10% by 2022 [2].

Among the major factors that are believed to be driving the growth of the market are the following [2]:

- rising demand for national security and safety;
- widening applications of spatial data in numerous industries;
- growing number of smart city projects and urbanisation;
- increased need for effective 3D GIS and land management, etc.

The market research states that the growing urbanisation along with rising demands for innovative smart city projects is the primary driving factors for the GIS market. GIS has also proven to be beneficial on stages of planning and implementation of many infrastructural projects for various government agencies.

According to Technavio report, in 2017, government end-user application sector dominated the GIS market [2]. Nobody can deny that every government must have access to full, relevant and detailed information regarding domains entrusted to it in real-time scale, and GIS is one of the most effective technologies able to provide it along with decision-making support tools. The GeoBuiz-18 report estimates the GIS and Spatial Analytics market to be the second largest after the GNSS and positioning market [3]. According to the report, the market is expected to grow from US$66.2 billion in 2017 to US$88.3 billion in 2020 growing at a CAGR of 12.4% [3].

Urbanisation and development of smart cities, integration of GIS-technologies into basic business analytic technologies, increasing import of GIS-solutions into transport sphere are also important factors affecting the growth of the market.

According to MarketsandMarkets's report, by the beginning of 2017, cartography represented the largest portion of the market [4]. Cartography improves the quality of decision making when selecting lands for agriculture, mining, disaster prevention, for city planning and transport infrastructure modelling. Consequently, demand for cartographic technologies steadily increases in fields of construction, creation and development of infrastructure, defence and security, transport, agriculture and forestry and environmental protection. City planning and development of smart cities also greatly influence the demand for cartography. Urbanisation and industrialisation, according to the report [4], stimulates growth of GIS market share in construction. Consequently, emerges a growing demand for observation equipment; for example, robotic measurement instruments, GIS and GNNS receivers and antennas. GIS databases are being widely applied for integration of geospatial data with CAD tools.

From the considered reports, we can conclude that modern GIS market is growing steadily, especially in smart city direction. Not all the demand is currently satisfied and smart cities' development and improvement always require new advanced technologies and ideas. Rostelecom has estimated that GIS market in Russia may grow up to 8.2 billion rub in 2020 [5].

Main analogues of CRISALIDE (CommunityViz, Cityworks, ArcGIS Online for State Government) are made in the USA and are mostly based on ArcGIS commercial technologies. In order for these products to operate in full capacity, the user is required to purchase at least a basic version of ArcGIS Desktop. The product "Cadastral office" is produced by ESTI MAP who is an official distributor of Pitney Bowes Software Inc. (a company from the USA that owns MapInfo, MapInfo Professional and other trademarks). This product requires the prior purchase of GIS MapInfo Pro and additional software products. Comparison of CRISALIDE project's features with domestic and foreign analogues is given in Table 1.

Currently, there are no CRISALIDE's analogues on the market that are fully manufactured in Russia and that could function in full capacity without any additional foreign software applications and that could successfully compete with foreign ones.

Therefore, we suggest to develop a system of city environmental management the purpose of which is to enhance the effectiveness of city environmental management and to reduce costs of city environmental planning at all management levels. Such system aims to integrate decision-making processes into the fields of strategic and territorial plans' creation, formation of urban policies, promotion of e-government, infrastructure and housing management (objects and public services, residential districts' regeneration), re-qualification of production sites and their development (renewal of ex-production sites, temporary use of abandoned buildings) and also land management in urban development regions.

Table 1 Comparison of features

Project's parameters	CommunityViz (City Explained, Inc., USA)	ArcGIS Online for State Government (ESRI, USA)	Cityworks (Azteca Systems, USA)	Cadastral office (ESTI MAP, Russia/USA)	CRISALIDE
1. Integration of maps of different formats	+	+	+	+	+
2. Scenario approach	+	–	–	–	+
3. 3D modelling	+	+	–	–	+
4. Decision-making support	+	–	–		+
5. Expert knowledge	–	–	–	–	+
6. Transport infrastructure	+	+	+	–	+
7. Ecological situation	+	+	–	–	+
8. Estimation decisions' of impact	+	–	–	–	+
9. Planning	+	+	–	–	+
10. Land management	+	+	+	+	+
11. Extension of functionality on demand	–	–	–	–	+
12. Cross-platform	–	–	–	–	+
13. Additional software required for operation	+	+		–	–

As core instruments and services of the CRISALIDE system, we propose to use scenario approach for situation modelling, expert systems for intelligent support of decision-making and for estimation of impact of the decisions on the situation development.

City environment management system will execute the following functions:

- collection, processing, storage and display of information obtained from different sensors for city environment monitoring, information about different city objects and related situations;
- access databases and knowledge bases in near real-time scale, perform ontological analysis of the city environment with regard to city infrastructure, housing and land management;
- analysis and prediction of dangerous situation development, of critical situations and potential threats that may influence the safety of life in the city environment (ecological, technological and road accidents, terrorist situations, etc.), provision risk estimation;
- development of recommendation for decision-makers (DM) in city management;
- documentation of the decision-making process, ensuring communication between DM and other users of the system in the process of decision making, estimation of the quality of decision-making support;
- model scenarios of complex spacial processes in the arbitrary time scale (e.g. re-qualification and development of production sites) and their visualisation on map;
- ensuring confidentiality, reliability, accuracy, credibility, communication of information and analysis of all factors that influence the management decisions;
- comfortable user interface and learning-by-doing.

3   Ontology Model of the City Environment

Ontology for city environment is formally a model of the city structure that includes components of city environment. Ontology of the city environment should include two interconnected ontologies: ontology of city environment as of a subject domain and scenario ontology that describes the behaviour of objects in the city environment.

The process of ontology creation for any information system includes the following steps:

1. identification of classes (concepts): basic notions from city environment domain;
2. setting up a hierarchy of classes (basic class → subclass);
3. identification of properties of classes: slots and their possible values;
4. filling the slots for class instances.

Basic requirements for the ontology of the city environment, according to [6], are the following:

1. simplicity: expressions and relations must be simple to use;
2. flexibility and scalability: adding of new notions and relations to the ontology must be clear and accessible;
3. universality: ontology must support different types of information (textual, graphical, etc.) and of knowledge about the city environment;
4. expressiveness: ontology must support description of the necessary number of attributes in order to adequately reflect the information and knowledge about the city environment.

Technologies for representation of knowledge about the city environment that is based on ontologies allow to speed up creation of new databases and to keep the existing ones up to date in accordance with information available.

Among the major components of city environment's ontology, we can highlight the following:

- information about constructions and buildings;
- information about transport system;
- information about energy system;
- information about environment;
- information about citizens;
- information about urban space use.

The modern approach based on object-oriented ontologies is used for representation of knowledge that constitutes the scenarios and the rule set. Development of scenario ontology includes the following steps:

1. Plotting scenario schemes for the simulated situations.
2. Plotting schemes for scenario phases.
3. Implementation of decision blocks into scenarios.
4. Implementation of concrete "elemental" actions from which the scenarios of simulated situations are built.

Therefore, the city environment ontology can be used as a basis for creation of decision-making support system for city environment management. Its distinctive feature is existence of scenario ontology that allows to model the most common activities in city management as well as unique ones. This technology is based on pre-designed rules that allow to automatically use contextual information about the city environment.

4 Design of the Intelligent Decision-Making Support System for City Management (Structure, Components)

The suggested city environment management system should be based on the following principles:

- scalability (retention of functionality with increase of the number of users and volume of processed information, possibility of further adaptation to the growing strain and additional functions);
- openness (possibility of increase of system's functionality);
- modularity (division of the system on independent modules with easy conjunction).

The term intelligent GIS is defined here as GIS that includes integrated tools and/or systems of AI [7]. The central part of the IGIS is a knowledge base that includes ontology of city environment. Ontology provides the 'framework' for representation concepts and for relations between them. Another part of knowledge base is storage of instances of real object from city environment.

Various active and passive sensors as well as information systems of global and local level can act as sources of information for intelligent decision-making support system for city management. In the process of obtaining data from different sources, the problems listed below may arise. Firstly, duplicated information may appear: information obtained from different sources but about the same city objects. However, similar data about the same object is not necessarily a disadvantage since it increases credibility of data and consequently increases quality of specific decisions. One of the means for solving the duplication problem and increasing the quality of data obtained from various sources is application of concept of integration, harmonisation and fusion of data [7].

Second problem is the issue of discrepancy between concepts in different systems. To format initial data to one standard, a unified information interoperability model on basis of ontology database was developed. Information interoperability model includes three ontology levels: domain ontology, geographical ontology and upper ontology.

The next component of intelligent decision-making support system for city management is GIS-interface. GIS-interface is a program component for visual representation of geospatial data in various digital formats and of objects stored in knowledge base. It combines different sources of geospatial data and program components that execute data processing using traditional methods.

GIS-interface allows:

- to update and display data in real-time mode along with processed results, predicted and modelled data;
- to display all infrastructure of observed area of urban activities;
- to set combinations of algorithms for execution of all stages of dangerous situation modelling (verification, interpolation prediction).

Library of mathematical functions is one of the important parts of intelligent decision-making support system for city management (IDMSSCM). Set of functions has to be open for access by any subsystem of IGIS, support changeability and expansion.

For example, for modelling spatial processes associated with dangerous situation development in urban activities location, following functions from library can be applied:

- mathematical model of different dangerous situations;
- search in location of rescue operation and etc.

To increase the quality of specific decisions made by user, it is necessary to include prognostic models in mathematical functions library, i.e. such models, that allow to obtain estimation of dangerous situation development in future instances of time, based on data obtained to the current point of time.

It should be noted that any function from mathematical functions library can be used in creating production rules for expert system.

Expert system (ES) is a system that uses expert knowledge (knowledge of specialists) to provide qualitative problem solving in the given subject area. Such systems can represent knowledge, have capability to interpret (research) its processes-reasoning and are intended for subject areas that require special abilities and a lot of experience.

Expert systems in IDMSSCM can be applied for solving of the following tasks:

- extraction of information from initial data;
- situation recognition;
- prediction;
- structural analysis of complex objects (e.g. districts, city areas, parts of terrain);
- diagnosing problems in IDMSSCM functioning and in connected technical complexes;
- configuring complex multiunit systems (e.g. distributed computer systems);
- planning execution sequence of operations leading to the chosen goal and etc.

It is important to note that this list is far from complete and is constantly extended along with development of geoinformation technologies and artificial intelligence (AI) methods.

Technological process of creation of the expert system demands particular form of interaction between its creator, also called as knowledge engineer, and one or several experts in the given subject area. Knowledge engineer "extracts" procedures, strategies and empirical rules from experts that they use for problem solving and "implements" this knowledge into the expert system.

Typical expert system integrated into IGIS contains the following major parts:

1. Knowledge base that contains rules (or knowledge in different forms of representation) that use this knowledge as basis for decision making.
2. Inference machine that realises process of solution search with use of knowledge from knowledge base.
3. User's interface that provides external interaction with IGIS.
4. Explanatory subsystem that informs the user about chosen decisions.
5. Knowledge base editor that allows to add, delete knowledge from knowledge base.

Aside from the listed components, the expert system, integrated in IDMSSCM, can contain other program means that are specified by the subject area.

Major and essential component of the expert system is the knowledge base. Creation of the knowledge base is a separate and reasonably complicated task.

IDMSSCM modelling system is intended for computer modelling of various spatial processes and also for visual generation of corresponding scenarios of processes development based on expert systems technology and represented as ontology database. Visual representation of the modelling allows the user to effectively estimate occurring process.

Modelling system allows us to solve the following tasks:

1. Building models of complex spatial processes based on their description in form of visual scenarios that are represented as two-dimensional digraph (block-scheme) where nodes are separate scenario tasks and decision-making points in which scenario branches on various execution routes depending on satisfaction of specified conditions. Scenario tasks can be executed both sequentially and in parallel, depending on block-scheme. For merging parallel branches, special nodes "connectors" are used. Scenario tasks consist of individual atomic actions and are as well represented as block-schemes connecting atomic actions. Task's actions can also be executed both sequentially and in parallel, depending on task's block-scheme.
2. Construction, debugging and testing of scenarios by field of study specialists mostly with use of visual drag-and-drop of icons, corresponding to tasks, solutions and actions, to the scenario and task scheme form and connecting them in accordance with scenario logic.
3. Scenario execution of complex spatial processes in optional time scale against digital map background, represented as moving simulated point objects with changing form, size, location, colour and transparency, etc. extended objects along with messages on natural professional language.
4. Interaction of number of complex processes, modelled on one as well as on several machines in network.
5. Manual object and process control option, modelling process in general (start/stop, time scale change, maps and etc.), scenario replay from control point and time jumps, etc.

5   Conclusion and Future Works

Implementation of CRISALIDE project will allow to solve the following tasks: to stimulate the development of e-government, to improve the quality of governmental management through creation and implementation of modern information technologies. In addition, further development and integration of the system for ministries of Russian cities may be sponsored from funds allocated on support of regional information technology projects.

There is a buoyant on such systems in Europe as well as in Russia which makes promotion of CRISALIDE project on European markets possible. European fund allocated on support of innovations in the field of information and communication technologies and in management may be used for funding the implementation of adaptable CRISALIDE system.

The second important industry for implementation of the developed system in construction is traditionally considered to be the driver of the economy. Stagnating housing market demands qualitative changes that will reflect the economical and demographical trends.

Increasingly relevant becomes the question about the most perspective areas and sites for residential and non-residential construction and their cost-benefit justification. The suggested decision-making system will allow to predict urban infrastructure development in the whole city and in its parts, to reveal deficits or surplus of construction and infrastructural objects, to consider social and economical aspects of regional development. These factors can influence the cost-benefit of construction objects in the conditions of surplus of demand and can influence predicted demand.

Moreover, the developed product may be of interest for investors and businessmen. Enterprises and organisations involved in any way in city environment (owners of retail, commercial areas, warehouses, invest funds, banks, private investors, etc.) may be interested in actual and illustrative information about perspectives of city environment and infrastructure development. The developed product will allow to not only manage the city environment but also to provide information, to plan and run scenarios of city development with consideration of business decisions of non-governmental companies and organisations.

Furthermore, CRISALIDE system allows to adapt a list of its functions to the subject domain and demands of the client. The system may be equipped with a set of management tasks, strategies and scenarios of development of specific industry, business, enterprise or organisation.

One more potential user of the CRISALIDE system is higher education institutions that train personnel for public service, ministries in the city environment and infrastructure domain. Also, CRISALIDE system can be adapted for higher education institutions' management.

References

Thematic investing. 21st century cities: global smart cities prime. Bank of America Merrill Lynch. https://static.esmartcity.es/media/2017/04/inversion-smart-cities-bank-of-america-merrill-lynch-negocio.pdf

Global GIS market analysis (2018) Technavio. https://www.technavio.com/report/global-gis-market-analysis-share-2018

Narain A (2018) The global GIS and spatial analytics market to touch US$88.3 billion by 2020. Geospatial World, May 2018. https://www.geospatialworld.net/blogs/gis-and-spatial-analytics-market/

Geographic Information system (GIS) market by component (hardware (GIS collector, total station, LIDAR, GNSS antenna) & software), function (mapping, surveying, telematics and navigation, location-based service), end user—global forecast to 2023. MarketsandMarkets https://www.marketsandmarkets.com/Market-Reports/geographic-information-system-market-55818039.html

Curishev E (2017) Data geography 2.0. "IT Weekly" Journal. https://www.it-weekly.ru/it-news/
    analytics/119063.html
Korpipää P, Mäntyjärvi J (2003) An ontology for mobile device sensor-based context awareness. In:
    Proceedings of CONTEXT, 2003, vol 2680 of lecture notes in computer science, pp 451–458
Popovich VV (ed) (2013) Intelligent geographic information systems. Nauka, Moscow, 324 pp (in
    Russian)

Develop a GIS-Based Context-Aware Sensors Network Deployment Algorithm to Optimize Sensor Coverage in an Urban Area

Meysam Argany and Mir-Abolfazl Mostafavi

Abstract Adequate coverage is an important issue in geosensor networks in order to fulfill the sensing applications in urban areas. GIS as well as optimization methods are widely used to distribute geosensors in the network to achieve the desired level of coverage. Most of the algorithms applied in urban domain suffer from the lack of considering real environmental information. In this paper, the problem of placing sensors to get optimum coverage is studied by investigating the concept of urban contextual information in sensor network. Then a local GIS-based context-aware framework of sensor network deployment optimization method is introduced. Obtained results of our algorithm under different working conditions and applications show the effectiveness of our approach.

Keywords Sensor network · Deployment · Optimization · Context-aware · Urban area

1 Introduction

In recent years, sensor networks have been increasingly used for different applications in smart cities ranging from urban environmental monitoring, tracking of moving objects, development of smart cities, and smart transportation system, etc. [1]. A sensor network usually consists of numerous wireless devices deployed in a region of interest inside the cities as well as other locations [2]. Despite the advances in the sensor network technology, the efficiency of a sensor network for collection and communication of the information may be constrained by the limitations of geosensors deployed in the network nodes. These restrictions may include sensing range,

M. Argany (✉)
Department of Remote Sensing and GIS, Faculty of Geography, University of Tehran, Tehran, Iran
e-mail: argany@ut.ac.ir

M.-A. Mostafavi
Department of Geomatics, Faculty of Forestry, Geography, and Geomatics, Laval University,
Quebec City, Canada
e-mail: mir-abolfazl.mostafavi@scg.ulaval.ca

© Springer Nature Switzerland AG 2020   179
V. Popovich et al. (eds.), *Information Fusion and Intelligent*
Geographic Information Systems, Advances in Geographic Information Science,
https://doi.org/10.1007/978-3-030-31608-2_13

battery power, connection ability, memory, and limited computation capabilities [3]. These limitations create challenging problems for the users of the sensor networks, which have pushed researchers from different disciplines in recent years to study various problems related to the design and deployments of efficient sensor networks in urban areas [4, 5]. Also, sensor networks have some limitations when it comes to the modeling, monitoring, and detecting urban environmental processes [5, 6]. Urban environmental elements like virtual and real obstacles, which exist in both static and dynamic natures, are also important to be considered in a realistic sensor networks deployment in urban areas. These restrictions usually affect the network coverage of the sensors. Spatial coverage of sensor networks has different definition according to different applications [3, 7–13]. Other examples of such elements include contextual information of the geosensors environment and physical urban phenomena in the network [14–16]. It is necessary to know how to use such information to make an appropriate and efficient sensor network deployment. For this purpose, one needs to introduce relevant models of the phenomena type, the accessibility or inaccessibility of the observation area, urban environmental conditions, spatial relations between sensors as well as urban deployment area, and related information availability. The complexity of urban area, as the sensing environment of sensor networks, with the presence of diverse obstacles may result in several uncovered areas in the sensing field [4]. Consequently, sensor placement affects how well a region of a city is covered by geosensors as well as the cost for constructing the network and creates its relations [5, 17, 18]. Hence, a fundamental issue in a sensor network deployed in an urban area is the optimization of its spatial coverage. Several optimization algorithms have been developed and applied in recent years to meet this criterion. Most of these algorithms often rely on oversimplified sensors, network models, and city model representations [6]. In addition, they do not consider environmental information such as city terrain models, man-made infrastructures, human and living creatures, and the presence of diverse urban obstacles in optimization process.

The impact of the quality of initial datasets used to deploy geosensors in the networks is another aspect of the complexity of wireless sensor network in urban areas [7]. Therefore, choosing the way of deploying sensors and the data accuracy needed to set up a sensor network in an optimal manner is difficult due to the abundance of available deployment algorithms as well as design of a consistent, reliable, and robust network. Thus, study of wireless sensor networks is a challenging task, as it requires multidisciplinary knowledge and expertise.

2 Methods

In order to develop a context-aware method for sensor network deployment, we need to precise the implication of the context and context awareness in this work. In order to provide a meaningful definition for the concept of "context" for sensor network deployment, the concept of sensor shall take the role of the main object of interest in such a definition. Here, the main object of interest is a sensor. Sensor network is considered as the object environment, which includes information on the sensors; for example, sensor position, orientation, and its spatial relations with other sensors in the network. Physical environment is composed of spatial objects in a given urban area in which the sensor is placed. It also may refer to the spatial relations among the objects, the specific locations in the urban environment such as desirable areas to be covered or restricted positions that are forbidden to set up the network. Here, context may also include information on sensor network preferences, objectives, and interests. Therefore, a comprehensive definition of context in sensor networks domain is proposed as follows: "Context is the whole situation, background, or environment of a sensor network. It includes information on sensor itself, the network, and the physical environment and their interactions in a given time."

That being said, for a context-aware sensor network deployment, we need to identify different contextual information (CI). Knowledge on CI for sensor results in awareness of any sensor on its position, orientation, and relation and interaction with other sensors and the physical environment and helps to hence decide on intelligent actions for sensor network optimization.

The contextual information may be very divers in their nature and require different strategies to be categorized in terms of integration in optimization algorithms. CI in sensor network deployment could be classified into spatial, temporal, and thematic information.

Spatial contextual information refers to the ability of defining objects positions and geometric relations. Spatial CI is not only about 2D or 3D position of sensors. A comprehensive framework of spatial contextual information may include sensors orientation, movement, routing, targeting, topology, and spatial dependencies and interactions. Hence, all information of spatial relations, interactions, proximity, and adjacency lies in this category.

Temporal contextual information concerns the temporal information and the temporal dependencies in data. Temporal information characterizes the dependency of a situation in the sensor network framework with the time and also indicates an instant or period during which some other CI is known or relevant. The objects and activities in the physical environment may change. For instance, position or attributes of an obstacle (e.g., its height) may change during a given period of time. A specific example of temporal CI is the information of a sensor movement and its trajectory in the network. Previous actions and movements of a sensor node may provide useful information for the next actions of current sensor or its neighbors.

Thematic contextual information in sensor networks constitutes the sensor specifications, network objectives, environment specifics, legal rules, etc. The information regarding the nodes names and roles, and their activity in the network is included in this category. Sensors activities may include measurement of the temperature, humidity, sound, or light. In terms of deployment, the type of sensor movement and its trajectory could be the sensor activity inside the network. Node name should be unique in the network in order to make it possible to be recognized and devolve its roles in multitasks networks. Sensor characteristics are sensor specifications, which have been designed during their manufacturing, e.g., their power supply, battery life, sensing range, temperature resistance, dimensions, input and output terminals, processing power, data storage capacity, send and receive information protocols, and etc. Network objective expresses the mission of sensor network to be fulfilled. This objective could be various in multitask networks. It may be varied from covering a whole, or a part of study area to monitor a phenomenon, or sensing different characteristics of the environment. Legal rules define specified terms and conditions for constructing and deploying the sensor networks, e.g., in which locations, sensor deployment is allowed or which parameters are permissible to be measured.

Based on mentioned issues on sensor networks deployment in urban areas, this paper presents a GIS-based approach to improve sensor deployment processes by integrating urban geospatial information and knowledge with optimization algorithms. To achieve this objective, the following approach that contains three specific parts is defined. First, a conceptual framework is proposed to show how geosensors and urban contextual information are integrated with a sensor network deployment processes. Then a local GIS-based context-aware optimization algorithm is developed based on the proposed framework. The extended approach is a generic local algorithm for sensor deployment, which accepts spatial, temporal, and thematic contextual information related to sensors and their distributing locations in different situations. Next, the accuracy assessment and error propagation analysis is conducted to determine the impact of the accuracy of contextual information on the proposed sensor network optimization method.

2.1   A Conceptual GIS-Based Context-Aware Framework for Sensor Network Deployment

Many parameters directly affect the sensing coverage; for example, topological relations among the sensors in the network, the interactions between the sensors and the environmental elements, and the relationship among the environmental elements themselves. Here, such information and relations are called network contextual information (CI). Specifically, CI defines the spatial dependencies between spatially adjacent nodes, nodes and urban obstacles, and obstacles themselves as well as the temporal dependencies between historical movements of nodes in the deployment process. Urban furniture, buildings, poles, etc. can be considered as the urban obstacles. The term of temporal information of sensor movements means the history of previous moves and trajectories of sensors which have had impact on the network coverage as well as their impact on the new probable moves. The so-called CI is used in the proposed framework to find good candidates positions of sensor nodes to fill uncovered areas and decide about the sensor's adequate actions in order to guide sensor network deployment.

The proposed conceptual context-aware framework consists of the following steps. First, the appropriate CI is extracted from the city area which is aimed to be studied. After introducing the CI to the framework, spatial and network databases are created. Spatial database contains CI related to the physical urban environment, while the network database comprises the CI belongs to sensors' configuration and relations. Accordingly, knowledge base is defined considering both databases. In the next step, a reasoning engine is applied using the predefined knowledge base. The optimization algorithm is also specified regarding the introduced local CI and tasks at hand. Afterward, the rules extracted from the reasoning engine along with the determined optimization method are applied to perform context-aware deployment actions. This may include a sensor move, delete, or insert. These actions may change the topology of the network, the configuration of the adjacent nodes, and consequently, the local coverage. As a result, local CI may be updated. Then the information in the knowledge base is changed and so on. This is an iterative algorithm and these actions are done until the desired level of deployment is achieved. The process of local optimization in the framework means that network configuration is changed locally at each step until the best coverage is obtained by considering the spatial, temporal, and thematic contextual information in the network (Fig. 1).

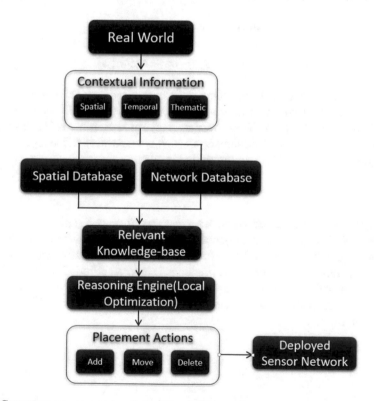

Fig. 1 Context-aware sensor network deployment framework

2.2 Local Context-Aware Optimization Algorithm for Sensor Network Deployment

According to the proposed framework for sensor network deployment, a local optimization algorithm is developed to tackle the sensors placement problem and maximize the spatial coverage of the network in an urban region. In the proposed algorithm, a set of sensors ($S = \{s_1, \ldots, s_n\}$) are randomly deployed in the first step. Then sensors start to move based on provided CI. There are some strategies for sensor movements to optimize the network coverage [19]. Next, the coverage(s_i) for each sensor is calculated. There are different definitions of coverage in literature. In this study, the blanket coverage is assumed. Blanket coverage requires placing a minimum number of nodes in an environment, such that every point of interest in the sensing area shall be adequately covered regarding tasks at hand [20]. Afterward, sensors are sorted in a priority queue, based on their coverage gain (g_i) obtained by considering different CI and following related moves in the network.

$$C_i = \text{coverage}(\text{move}(s_i, S)), \quad \forall s_i \in S \tag{1}$$

$$g_i = \text{coverage}(\text{move}(s_i, S)) - \text{coverage}(s_i, S), \quad \forall s_i \in S \tag{2}$$

Then the sensor with the maximum gain is selected, which is the sensor that obtained the highest coverage improvement by its movement in the network and stands at the top of the queue. The movement types of sensors are related to the local CI as well as sensor network mission. By changing the position of the topmost sensor of the queue (s'_u), the network configuration is updated.

$$s'_u = \text{move}(s_u, S) \tag{3}$$

Next, the coverage gain of the adjacent sensors of moved sensor is recalculated and their ordering in the priority queue is updated. In the next iteration, the (new) topmost sensor of the queue is chosen to move and so on. This optimization process is conducted iteratively until one of the predefined stopping criteria is reached (Fig. 2). The iteration will be stopped when topmost sensor movement does not improve overall coverage of the network. The maximum gain will be defined based on the deployment mission.

Fig. 2 The pseudocode of
the local context-aware
algorithm

```
1: Initial sensor positions
2: Define sensor moves based on CI
3: Compute potential gains (gᵢ)
4: Make the priority queue
5: while max gᵢ > ε do
     6: Select sensor with highest gᵢ
     7: Compute new position for
        selected sensor
     8: Move selected sensor
     9: Determine set of neighbors
     10: Update potential gain for
        neighbors
11: end while
```

2.3 Impact of the Spatial Data Quality on Sensor Network Deployment

The sensor placement optimization algorithms that are applied in our experimentation use urban spatial information to calculate spatial coverage. This way, visible and invisible objects are identified, and hence, covered and uncovered areas in the urban region of interest are defined. The quality of spatial data has a direct impact on the estimation of these values. Among different data quality elements, positional accuracy and completeness were selected to study because of their direct impact on the estimation of the visibility [7]. The positional accuracy may be presented as a small displacement in the position of the objects, which can be either horizontal or vertical or both. Even a few centimeters inaccuracy in horizontal or vertical positions of objects or sensors can block the line of sight between a sensor and a target. Same reasoning may be applied for incompleteness of databases.

3 Experimentations and Results

3.1 Study Area and Datasets

The study area of the experiments is a part of the campus of Université Laval, located in Quebec City, with a dimension of 300 m by 300 m (Fig. 3). For the experiments, 12 sensors have been assumed to be deploying as surveillance cameras with 360° horizontal, ±90° vertical sensing angle, and 50 m of effective sensing range.

In order to deploy sensors using the proposed context-aware algorithm, unlike many previous optimization algorithms which do not consider the environment topography or real obstacles, the sensors were deployed over the terrain model of study area. The buildings and other urban features and obstacles have been added

Fig. 3 Part of the campus of
Université Laval

over the terrain model containing their horizontal shapes and coordinates as well
as their heights (Fig. 3). The test site contains the buildings of library, Faculty of
Educational Sciences, Faculty of Philosophy, two main streets, and a parking lot.

3.2 Different Contextual Information Situations

Terrain Surface and City Features Model:

The first category of CI is the city surface model of the sensor network distributing
area, besides the information of the network. Having this information the height
of city objects, and the terrain elevation model in study area is provided, and as
a result, the obstacles which bared the sensing field of the sensors are introduced.
Accordingly, after running the context-aware algorithm, the sensors start to move
and the network configuration is changed based on provided CI. In each iteration,
sensors find their best positions, over the buildings or on the terrain surface, based
on the coverage improvement they might provide.

Figure 4 depicts the coverage improvement during running the optimization algorithm for one of runs which returned the best coverage over 32 runs. It shows that in

Fig. 4 Coverage
improvement over iterations
for run number 18 which
returned best results for the
context-aware method

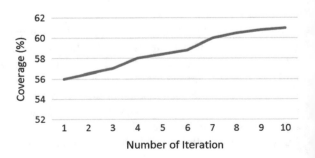

Table 1 Results of context-aware algorithm compared to CMA-ES optimization algorithm

Method	Average coverage (%)	Best coverage (%)
Context-aware	64.14	66.26
CMA-ES	61.76	64.44

each iteration, the overall network coverage is improved, while the optimum coverage in this network is reached at just 10 iterations, which is significantly less than other deployment optimization methods with the same initial parameters. For instance, the deployment progress was compared to the CMA-ES [11] optimization algorithm. In order to avoid the impact of initial positions over the final results [21], 32 runs for each method with 32 different initial random positions were performed. The results are presented in Table 1.

Urban Legal Restrictions:

Urban legal restrictions are considered as the next category of CI which was introduced to the sensor network optimization context-aware algorithm. In this study, inside the streets and parking areas (dark green area in Fig. 3) and top of two towers (red towers in Fig. 3) were assumed to be unauthorized places for sensor installation. Considering restricted areas in context-aware optimization, sensor action is changed, and new moves are defined within the optimization process. Again, 32 runs with different random initial positions were applied to avoid unexpected impact of initial sensor positions. Table 2 shows the result of five best runs.

Desirability of Coverage:

Desirability of coverage is another type of urban thematic CI which is introduced to be considered in the proposed context-aware optimization process. Suppose that there are some places on the campus, where sensors should not be installed, there is a high interest in those regions to be covered. In this study, "Avenue des Sciences Humaines," the street between towers and the library building was introduced as the area with high interest to be covered, while it is unauthorized zone for sensors to be deployed. Table 3 presents results for both coverage of interested zone plus overall coverage for the best run of algorithm as well as the average result over 32 runs.

Evaluation the Minimum Number of Sensors:

To evaluate the investigation of minimum number of sensors needed in each sensor group network to cover a region of interest in an urban area, we studied the sensor

Table 2 Results of five best runs for context-aware algorithm considering urban restrictions as CI

Run no.	Overall coverage (%)	Number of iterations
21	61.78	8
26	61.59	7
20	60.87	9
14	59.16	8
5	58.42	7

Table 3 Results of context-aware algorithm using desirability of coverage in an urban region

	Average coverage (%)	Best coverage (%)
Coverage on "Avenue des Sciences Humaines"	85.07	86.55
Overall coverage	63.51	65.83

deployment over the campus using four groups of 8, 12, 16, and 20 sensors in the network. The results show that approximately 50% of sensors are applied to cover the region of interest when the number of sensors is 8 and 12. While in the group of 16 and 20, there are still six sensors which are used to cover the "Avenue des Sciences Humaines." This means that the targeted street is covered using maximum of six sensors, and the average coverage 85.07% reaches by adding just two sensors from the previous test, which had four sensors to be deployed. In next tests, by considering 16 and 20 sensors, the overall coverage over the whole study area is improved, while the coverage on "Avenue des Sciences Humaines" remains unchanged due to reach the optimum number of sensors to cover this area. This confirms the capability of proposed method to optimize not only the overall and interested coverage of the network but also the number of sensors used to perform desired actions. Table 4 has the result of this investigation.

Finally, to investigate the impact of positional accuracy and completeness of the dataset on the spatial coverage of a sensor network, we prepared five maps with different resolutions for the campus area, which contain the terrain surface information plus the height of urban features like buildings and other obstacles. The resolution variation is from 500 cm (low resolution) to 50 cm (high resolution) and a map with 10 cm resolution is considered as the ground truth dataset to validate the results. All maps are from the same area, which previous tests were done. The experimentation consists of deploying eight sensors considering the first category of CI (terrain and urban surface model) inside the study area. Similar to previous tests, the proposed context-aware optimization algorithm was run 32 times over the study area in order to reduce the probable impact of the initial sensor position on final results. Then for each map, average coverage and best coverage of 32 runs were calculated. Table 5 shows the results. The results show that both average and best coverage are improved by increasing the data resolution. The proposed algorithm performs better in higher resolution because the urban features and obstacles are appeared with more details. In return, when the optimum sensors positions have been found from different testing resolutions are applied on the 10 cm reference map, the average and best coverage are reduced. This could be interpreted with the same reason. When the context-aware algorithm finds optimum sensor positions using a lower resolution data, then we applied same positions over a higher resolution map, we considered more details of a higher resolution map to calculate coverage using positions that are obtained based on lower data resolutions.

Table 4 Results of increasing number of sensors to cover the region of interest

	Total number of sensors	Number of sensors to cover RoI	Average coverage (%)	Best coverage (%)
Coverage on "Avenue des Sciences Humaines"	8	4	69.36	71.10
Overall coverage			51.67	53.22
Coverage on "Avenue des Sciences Humaines"	12	6	85.07	86.55
Overall coverage			63.51	65.83
Coverage on "Avenue des Sciences Humaines"	16	6	85.07	86.55
Overall coverage			65.29	68.04
Coverage on "Avenue des Sciences Humaines"	20	6	85.07	86.55
Overall coverage			66.86	69.62

Table 5 Results obtained from the context-aware method, considering the urban surface model

Resolution (cm)	Average coverage (%)	Best coverage (%)	Best coverage from best configuration over 10 cm resolution (%)	Average coverage over 10 cm resolution (%)
500	48.61	51.36	48.39	45.92
300	48.85	51.73	47.42	46.28
200	50.39	52.96	51.60	50.05
100	51.67	53.22	51.68	49.85
50	53.85	55.59	53.20	51.69

4   Conclusions

The purpose of this paper was certainly not to overcomplicate the optimization process, but rather to find a flexible methodology that can locally accommodate all relevant urban information that would have an impact on sensor placement. To do so, a local optimization framework was introduced. The extended optimization algorithm can come up with different sensor placement configuration according to the various circumstances, environmental information, and/or sensor parameters encountered. Consequently, if there are any changes in sensor parameters or urban environment, the context-aware algorithm can simply take in new contextual inputs and regenerate a new sensor placement design adapted to the new situation. The outstanding advantage of the proposed context-aware algorithm was that it was designed independently of any specific CI. Thus, it is able to take into consideration different types of information based on specific network applications, urban requirements, and tasks at hand.

References

1. Nittel S (2009) A survey of geosensor networks: advances in dynamic environmental monitoring. Sensors (Basel) 9(7):5664–5678
2. Lewis FL (2004) Wireless sensor networks. Smart Environ Technol Protoc Appl 1–18
3. Ghosh A, Das SK (2008) Coverage and connectivity issues in wireless sensor networks: a survey. Pervasive Mob Comput 4(3):303–334
4. Ghosh A (2004) Estimating coverage holes and enhancing coverage in mixed sensor networks. In: 29th annual IEEE international conference on local computer networks, pp 68–76
5. Thai MT, Wang F, Du DH, Jia X (2008) Coverage problems in wireless sensor networks: designs and analysis. Int J Sens Netw 3(3):191–203
6. Li BM, Li Z, Vasilakos AV (2013) A survey on topology control in wireless sensor networks: taxonomy, comparative study, and open issues, vol 101, no 12
7. Aziz N, Aziz K, Ismail W (2009) Coverage strategies for wireless sensor networks. World Acad Sci Eng Technol 50:145–150
8. Ahmed N, Kanhere S, Jha S (2005) The holes problem in wireless sensor networks: a survey. ACM SIGMOBILE Mob Comput Commun Rev 1(2):1–14
9. Huang C, Tseng Y (2005) A survey of solutions to the coverage problems in wireless sensor networks. J Internet Technol 1:1–9
10. Adriaens J, Megerian S, Potkonjak M (2006) Optimal worst-case coverage of directional field-of-view sensor networks. In: 2006 3rd annual IEEE communications society on sensor and ad hoc communications and networks, pp 336–345
11. Akbarzadeh V, Gagne C, Parizeau M, Argany M, Mostafavi MA (2013) Probabilistic sensing model for sensor placement optimization based on line-of-sight coverage. IEEE Trans Instrum Meas 62(2):293–303
12. Argany M, Mostafavi MA, Karimipour F, Gagné C (2011) A GIS based wireless sensor network coverage estimation and optimization: a Voronoi approach. Trans Comput Sci XIV 6970:151–172
13. Paul AK, Sato T (2017) Localization in wireless sensor networks: a survey on algorithms, measurement techniques, applications and challenges. J Sens Actuator Netw 6(4)
14. Sun JZ, Sauvola J (2002) Towards a conceptual model for context-aware adaptive services. In: Proceedings of 8th international scientific and practical conference of students, post-graduates

and young scientists modern technique and technologies MTT'2002 (Cat. No.02EX550), pp 90–94

15. Park S, Savvides A, Srivastava M (2000) SensorSim: a simulation framework for sensor networks. In: Proceedings of the 3rd ACM international workshop on modeling, analysis and simulation of wireless and mobile systems, pp 104–111
16. Mahfouz AMA, Hancke GP (2018) Localised information fusion techniques for location discovery in wireless sensor networks. Int J Sens Netw 26(1):12
17. Shit RC, Sharma S, Puthal D, Zomaya AY (2018) Location of Things (LoT): a review and taxonomy of sensors localization in IoT infrastructure. IEEE Commun Surv Tutorials 20(3):2028–2061
18. Bharathi Priya C, Sivakumar S (2018) A survey on localization techniques in wireless sensor networks. Int J Eng Technol 7(1.3):125
19. Karimipour F, Argany M, Mostafavi MA (2014) Spatial coverage estimation and optimization in geosensor networks deployment. In: Ibrahiem MM, Ramakrishnan S (eds) Wireless sensor networks, from theory to applications. CRC Press, Taylor & Francis Group, pp 59–83
20. Fan G, Jin S (2010) Coverage problem in wireless sensor network: a survey. J Netw 5(9):1033–1040
21. Argany M, Mostafavi MA, Akbarzadeh V, Gagne C, Yaagoubi R (2012) Impact of the quality of spatial 3D city models on sensor networks. Geomatica 66(4):291–305

Printed in the United States
By Bookmasters